TRANSFER OF PLANT PROTECTION TECHNOLOGY AMONG DRYLAND FARMERS

The Author

Dr. P.S. Doijad is communication specialist in the Directorate of Rice Research, Rajendranagar, Hyderabad. He obtained M.Sc., (Agri. Extension) and Ph.D from Punjabrao Krishi Vidyapeeth, Akola and B.Ed. from Nagpur University. He worked as Technical Officer in the National Plant Protection Training Institute, Hyderabad and Lecturer, Audio-Visual Aids in the Extension Education Institute, Nilokheri (Haryana) before joining Directorate of Rice Research Hyderabad. He has wide experience of 22 years in the field of training, research and extension. He is a Co-author of a book *Audio-visual Aids* with Dr. S.V. Supe apart from his contribution of several research papers and popular articles in various journals.

Transfer of Plant Protection Technology Among Dryland Farmers

P.S. Doijad

Mittal Publications
NEW DELHI-110059 [INDIA]

Publication of this volume is financially supported by the Indian Council of Social Sciences Research (ICSSR), New Delhi. The responsibility of the facts stated and conclusions reached is entirely that of author; and not of the Council.

First Edition 1998

ISBN 81-7099-698-8

Published and printed by K.M. Rai Mittal for Mittal Publications,
A-110, Mohan Garden, New Delhi-110059, India.

Phones : 5648028, 5648725
Fax : 91-11-5648733 *Gram* : Mittalbux

SHOWROOM

4594/9, Daryaganj,
New Delhi - 110002

Phone : 3250398

ACKNOWLEDGEMENTS

With profound sense, I express my heartiest gratitude to Dr. S.V. Supe, M.A. (Eco.) M.Sc. (Agri.), Ph.D. (IARI), Research Guide and Head, Department of Extension, Post-Graduate Institute, Punjabrao, Krishi Vidyapeeth, Akola for his constant inspiration, benevolent guidance, instructive suggestions and painstaking supervision through the course of this research with thoroughness which helping in bringing it to the present form.

I feel myself fortunate to express my gratefulness to Dr. P.D. Wangikar, Associate Dean, PGI and Dr. P.W. Amin, Vice-Chancellor, Punjabrao Krishi Vidyapeeth, Akola for providing affectionate guidance, tender care and facilities required during the course of investigation.

I am immensely thankful to Dr. D. Sen, Director (Extension and Transfer of Technology), National Institute of Rural Development for valuable suggestions, inspiring advice at various stages of the investigation which helped in the completion for this research.

I feel greatly indebted to Dr. Nagarcenkar, Director, NDRI, Karnal for extending the computer facilities for analysis for data.

The active co-operation of the officials of the Panchayat Samiti, Kuhi and other concerned departments helped in fulfilling my information requirements. I acknowledge their help.

The main contributions to my data bank were the farmers of Panchayat Samiti, Kuhi who were selected as respondents for this investigation. I cannot forget their cooperation and owe a great sense of gratitude to them who fed all necessary informations happily and smilingly for the study.

A choice of thanks to Shri P.S. Shinde, Dr. R.S. Bhopale, Assistant Professor in Agri. Extension, P.G.I., Punjabrao Krishi Vidyapeeth, Akola, Dr. A. P. Purandare, Research Associate, NIRD, Hyderabad and Mr. N.R. Kude, Sr. Research Assistant, P.K.V., Akola for providing necessary help during the course of investigation.

I am fully aware that my concepts regarding plant protection and transfer of technology have been flourished by galaxy of scientists contributing over years. I am immensely indebted to all to them.

My special thanks go to my wife Mrs. Sushila Doijad, son Yawdas and daughters Alka and Sunanda who shouldered all the household responsibilities patiently during the period of this investigation.

Finally, I am thankful to those all who directly or indirectly rendered all sorts of assistance for the completion of this investigation.

P.S. DOIJAD

CONTENTS

LIST OF FIGURES

LIST OF TABLES

1

INTRODUCTION

The earlier research efforts on dryland agriculture was started by Tamhane in 1923 at Manjri in Maharashtra. Subsequently a rather comprehensive programme was evolved by Kantikar for improving the productivity of rainfed farming. This formed the basis for the dry farming scheme started in 1933 by the Indian Council of Agricultural Research at Solapur, Rohtak, Bijapur, Raichur and Hagari. The outcome of these practices commonly known as Bombay, Hyderabad and Madras dry farming practices. The basic features of the dry farming practices developed in the past were bunding to conserve soil moisture, application of farm yard manure to supply plant nutrients, deep ploughing once in three years, shallow preparatory cultivation and interculture, low seed rate and wide spacing of crops.

Later in 1954, soil conservation research and demonstration centres were established in different parts of the country. It is experience of this kind that led to the intensification of dry farming research in 1970 through the initiation of 23 All-India Coordinated Research Projects on dryland agriculture with the collaboration of the Government of Canada. Meanwhile, International interest in upgrading the productivity as well as the stability of production of the dry farming region of Asia, Africa and Latin

America grew and resulted in the establishment of the consultative group for International Crop Research Institute for the Semi Arid Tropic at Hyderabad in July 1972. In the year 1987, the Coordinating Cell and Research Centre of Hyderabad on Dryland Agriculture has been changed into Central Research Institute for Dryland Agriculture which also coordinate the research activities of all 23 research centres located throughout the country established with the objective to test the research immediately on area basis for immediate. Transfer of Technology to vast areas of drylands in the country.

In order to meet the projected needs of 250 millions tonnes in the year 2000 A.D., it would be absolutely necessary to increase crop production in the dryland areas to feed up 100 crores, the expected population of our country by 2000 A.D., as against today's about 80 crores on March 1988, as signalled by the demographers[62].

It is also a fact that the impact of green revolution in India has been restricted more in the irrigated areas leaving aside the dry farming areas which not only make important contributions to the National food basket including the important protein source, pulses, but also provide much of the industrial raw materials like cotton and oilseeds. Today, these areas are contributing about 45 per cent each of cereals and 75 per cent each of pulses and oilseeds[98].

The system of crop cultivation in low rainfall regions receiving 375-1125 mm annual precipitation and has less than 30 per cent of sown area under irrigation refers as dry farming. As many as 115 districts fall under this category spread over 75 per cent of the cultivated area of the country. About 280 million people are living in this belt and comprise of small and marginal farmers. The only way to meet the shortage of pulses, edible oils and cotton is to improve the productivity of these crops in the dryland areas of the country. Increased agricultural production is possible by Transfer of Technologies at the farmers' level in these areas, which constitute a large belt of the country.

1.1 NEED OF STUDY

The phrase 'Transfer of Technology' has recently come to be used in agricultural development. Transfer of technology refers to the process by which recommended practices, produced by research and development, are transmitted through extension agents to the farmers[11]. Among the several facets of such technologies, plant protection occupies a vital position. Food and Agriculture organisation[19] has estimated that as much of 40 per cent of the Food produced by the farmers, is infact eaten by the pests. The annual crop losses due to pests in our country ranged between 20 to 25 per cent of yield, which came to about Rs. 6,000 crores in terms of money[66].

Mann[42] stated that there is no point in raising the potential for increased crop yield through breeding improved varieties, only to waste it by ignoring crop protection. Therefore, it is not enough that we produce, but what we produce must also be protected. Plant protection should, therefore, be regarded as "Investment Protection". A survey by the National Council of Applied Economic Research revealed that every rupee spent on plant protection, other things being constant, would protect on an average crop output worth Rs. 4 which otherwise would have been lost.

Although the use of pesticides is expected to reach 75,000 tonnes of technical grade material by 1989-90 at the end of Nations Seventh Development Plan[81]; the use of pesticides has not much progressed in India as compared to other developed countries. We are using about 400 grams technical grade per hectare on an average as compared to 10,790 grams by Japan, 1,490 grams by USA and 1,870 grams by Europe[6]. Hence, there is a need to pay more attention on plant protection technology for minimising the losses caused by insect pests and diseases.

The success of such a sophisticated technology depends largely on its transfer at the farmers' level who are its ultimate users. Effective transfer of any technology in the farmers' fields necessitates a strong research system to develop the technology relevant to the farmers' needs, problems, resource conditions; a strong extension machinery equiped with communication skill and ably supported by needed support facilities; coupled with several kinds of personal, social, economical, psychological, situational and communicational characteristics in the farmers' system, so that farmers can adopt it in their situation. Variation in the characteristics of each of these systems will result in different kinds of farmers' response in terms of adoption of modern technologies vis-a-vis their agricultural productivity.

A close observation of the pest control efforts lead to a series of actions on the part of the farmers which can be called as the plant protection behaviour of the farmers. Before making the efforts to accelerate the plant protection; it is necessary to have a clear picture of the plant protection status of farmers which includes knowledge, skill, attitude and adoption[51].

Gartrell and Gartrell[21] pointed out that the social status is the main cause for variation in the adoption behaviour of the farmers. Sen and Doijad[73] supported this view and stated that the sophisticated technology like plant protection requires favourable socio-economic condition of the farmers for bringing about their increased adoption.

Jaiswal and Purandare[30] observed that dryland farmers were mostly confined to the use of improved seeds and fertilizers and not plant protection practices. Even after attempting to motivate the farmers to adopt plant protection measures by conducting demonstration, imparting training to farmers, organising visits to dryland research stations, the farmers' responses were very poor. The reasons for such poor responses with reference to the plant protection measures in dryland areas need to be analysed in terms of various problems in the transfer of technology among the dryland farmers.

Few attempts are made to study the present topic of investigation, particularly in the dryland areas. However, few studies, Government of Bihar[24], Evaluation Committee on Lab to Land Programme[2], Kunnal *et al.*,[39] had been conducted as part of their investigation, highlighting the different problems of Transfer of Technology of plant protection at the farmers' level adoption. In view of this, the present study has been conducted with the following specific objectives.

1.2 OBJECTIVES

1. To study the socio-personal, socio-economical, situational psychological and communicational characteristics of the dryland farmers.
2. To study the adoption and re-invention of plant protection measures by the dryland farmers.
3. To study the relationship between plant protection status and characteristics of the dryland farmers.
4. To study the factors affecting plant protection status of the dryland farmers.
5. To study the problems of transfer of technology in plant protection adoption by the dryland farmers.

1.3 FORMULATION OF NULL HYPOTHESES FOR THE STUDY

1. There is no relationship between the plant protection status and the characteristics of dryland farmers.
2. The dryland farmers do not differ from other farmers in adoption and re-invention of plant protection measures.
3. There are no factors affecting plant protection status of dryland farmers.
4. There are no problems in transfer of technology in plant protection adoption to dryland farmers.

2

REVIEW OF LITERATURE

Before taking any research investigation it is of prime importance to take a review of the work done by previous researchers. The project has got different components namely, characteristics of dryland farmers, adoption and re-invention of plant protection measures, plant protection status along with its affecting factors and problems of transfer of technology. The review of literature is therefore arranged under these heads. Besides these, conceptual model and definitions of the terms used in the present investigation have also been given in this chapter.

2.1 CHARACTERISTICS OF DRYLAND FARMERS

Janakiramaraju[31] observed that the farmers in irrigated areas were superior to those in non-irrigated area regarding farm size and socio- economic status. However, they did not differ in age, education, social participation and farming experience. Subramanyan and Sripaul[90] in their comparative study on "Garden land and dryland" farmers reported that the dryland farmers were more aged, less educated, had small farm size and less social participation. Vijayaraghavan[101] stated that participants and non-participants in dryland agricultural project differed significantly in respect of their education, social participation, economic status and risk preference, whereas they were not different in their age, occupation, farming experience and farm size. Reddy and Reddy[70] stated that the majority of the dryland farmers were distributed to low income level with regard to jowar and castor crop cultivation.

An overview of the above cited literature reveals that the dryland farmers were low educated, had small farm size, low social participation and mostly engaged in subsidiary occupations having low income level.

Murthy[89] reported that indiscriminate use of folidal spray even for controlling diseases were observed among vegetable growers of Bangalore district due to lack of knowledge about plant protection measures. Sinha *et al.,*[84] found that higher percentage of small and marginal farmers had lack in knowledge in the area of plant protection. They asked for intensive training in plant protection. Mayani and Sheth[44] found the knowledge of the farmers about plant protection was poor, it was fair in agronomic and manurial practices. Reddy and Reddy[70] stated that the majority of the farmers of the dryland agriculture under study were distributed in an average knowledge category about improved agricultural practices.

The above studies indicate that knowledge of the farmers is low about plant protection.

Diffusion researchers had shown that the farmers use different sources for getting agricultural information but perceive few of them as credible. There may be several reasons for this kind of differential information seeking habit of the farmers. However, the authenticity and specialisation of an information source largely decide the acceptance or rejection of an innovation advocated by that source[77]. Many information sources operate at the village level, which help in communicating ideas to the farmers. They have been classified as Institutionalised and Non-institutionalised sources by Jha and Singh[32] and formal and informal sources of information by Rao and Moulik[63]. Among these several sources it is observed that the use of personal sources of information is common at every stage of the adoption process[76]. Rao *et al.,*[3] found a tendency on the part of the farmers of non-innovative village to seek information from their leisure associates in the matter of cultivation. They also observed higher level of adoption in the headquarter villages of extension workers. This is attributed to more frequent utilisation of change agents as a source of information by the farmers in the headquarter villages as compared to the remote villages. Mruthyunjayam[47] stated that, at awareness stages regarding tobacco cultivation, salesman and Government agency played important role in the use of fertilizers and plant protection measures; whereas, in the stages like interest, trial and adoption, the neighbours and personal experience of the farmers were considered as important sources of information. The above reviews stated that though many sources of information operate at the village level, farmers used to utilize only the sources of personal nature with regard to different farm practices.

2.2 ADOPTION AND REINVENTION

Sundaraswamy[93] concluded that application of farm yard manure, use of certified seed and proper seed rate were adopted by all the respondents, while 50 per cent did not adopt the practices like plant protection measures, proper spacing, and application of fertilizers recommended for jowar cultivation. Somasundaram[88] reported that there was a wide variation in the extent of adoption of all the recommended practices except seed rate by small farmers. Desai[13] revealed that majority of participant and non-participants farmers adopted correctly the practices demonstrated in Block Demonstration; namely, interculture, seed rate and sowing time. The complex practices like plant protection, fertilizers and manures were partially adopted by both the categories. Venkiduswamy[100] also reported that, in general, adoption was low in plant protection, application of fertilizers and irrigation. Dhankumar[14] reported that about one fourth of the respondents adopted the plant protection measures in full, while one third of them did not adopt at all. Similar findings were reported by Raje[58]. He observed that the plant protection measures were not followed by the farmers in dryland agriculture.

Until very recently we assumed that adoption of innovation meant the exact copying or imitation of how innovation had been used previously in a different setting. Invention is the process by which a new idea is discovered or created, while adoption is a decision to make full use of an innovation as the best course of action available. Thus, the difference between invention and adoption however is not so clear cut when we acknowledge that an innovation is not necessarily a fixed entity as it diffuses within a social system. For this reason, reinvention seems like a rather appropriate word to describe the degree to which an innovation is changed or modified by the users in the process of its adoption and implementation stage for certain innovations and for certain adopters[71].

Further, Rogers[71] stated that reinvention is not necessarily bad. Reinvention can be beneficial to the adopters of innovation. Flexibility is the process of adopting an innovation which may reduce mistakes and encourage customisation of the innovation to fit it more appropriately to local and/or changing conditions. Some of the reasons for reinvention exist in the innovation itself while others involve the individual or organisation that is adopting new idea. Innovation that are relatively more complex, lack of detailed knowledge about the innovation to learn may lead to reinvention. When innovation is implemented in order to solve a wide range of users' problems in such cases also reinvention occurs.

Singh *et al.,*[86] found that the farmers did not adopt an innovation as such. They observed the cases of reinvention with regard to change in the recommended seed rate of high yielding variety of rice. Poor knowledge of innovation and inappropriate innovation as perceived by the farmers were the causes for reinvention. Jha and Shaktawat[33] inferred that high yielding variety programme of *bajara* had been a partial success. As a testimony, they reported that 88 per cent of *bajara* growers under study, did not adopt full improved practices. Chauhan *et al.,*[10] and Sharma and Nair[75] found that the small farmers used the recommended modern inputs such as plant protection chemicals and fertilizers much below the recommended dose. Similar findings were reported by Padheria and Patel[52]. They pointed out that most of the farmers used below recommended dose of costly inputs like fertilizers, manures and plant protection chemicals on the cotton crop.

From the foregoing review of the past researchers it could be concluded that the adoption of the simple recommended practices for crops in dryland agriculture by the farmers was comparatively more than the complex practices like application of chemical fertilizers and plant protection measures. Moreover, the adoption of such recommended practices was found to be less than the recommendations, thereby implying the occurrence of reinvention in the use of costly inputs.

2.3 PLANT PROTECTION STATUS

The plant protection status allows us to know as to how much knowledge the farmer possesses about scientific plant protection practices, his attitude towards plant protection, his skill in adoption of the practices in controlling the pests and actual adoption of plant protection measures[51]. This being the new concepts of studying the plant protection status at the comprehensive level rather than separate study of the different components, no direct reference could be made available. However, attempt has been made to review the relevant researches on different components of plant protection status of the farmers along with their affecting factors. Different researchers studied the different components of plant protection status independently.

2.3.1 Knowledge of Improved Practices

Sukumaran[92] and Gopalakrishna[23] found that age is not correlated with the knowledge level of the farmers. In contradiction with this, Nagendranath[48] observed the significant relationship of age with the knowledge level of the farmers. Narayan Rao[49] reported that there was a positive association between formal education of cultivators and their knowledge. Sukumaran[92] concluded that as the level of education increased knowledge also increased. Gopalakrishna[23] and Nagendranath[48] observed that farmers with higher education tend to have more knowledge than those with lower educational levels. Dhanakumar[14] observed that educational status of sorghum growing farmers was found to have significant as well as positive association with their knowledge level.

Narayana Rao[49] reported that the farmers with large size of land holding were getting more knowledge of improved practices. Sinha[83] found that the farmers with medium size of land holding were better disposed to gain new knowledge on farming. Janakiramraju[31] reported that farm size had significant and positive association with the knowledge level of farmers. Nagendranath[48] reported that small farmers possessed higher knowledge than marginal farmers.

Vittal[102] stated that there was negative relationship between social participation and gain in knowledge among trained farmers. Nagendranath[48] reported that the farmers who had social participation were having better knowledge than those who had no social participation.

Trigunayat[96] found that there was positive association between socio-economic status and knowledge of the farmers. Gopalakrishna[23] found a positive and significant correlation between knowledge and socio-economic status. Janakiramaraju[31] reported that socio-economic status had significant and positive association with knowledge level of farmers. Dhanakumar[14] found a positive and significant association between knowledge and economic status.

Janakiramaraju[31] reported that economic orientation had non-significant and positive relationship with knowledge level of dryland farmers. However, positive and significant correlation between economic motivation and knowledge of farmers was observed in the past studies by the different researchers namely Dhanakumar[14], Kulhari[38], Pandey[50] and Pathak[53].

A positive correlation between risk preference and

knowledge of the farmers was reported by Pandey[50] and Kulhari[38]. Dhanakumar[14] also reported similar findings between risk orientation of jowar cultivators and their level of knowledge. Non-significant relationship between these two variables was observed by Pathak[53] and Pathak and Mujumdar[54]. However, Reddy and Reddy[70] observed positive and significant correlation with the risk preference and knowledge level of farmers.

Janakiramaraju[31] reported that the scientific orientation had significant and positive association with knowledge level of farmers. This view was supported by Mannivannan[43].

Singh[79] found that institutionally and non-institutionally trained farmers significantly differed from untrained farmers with respect to knowledge of package of practices of high yielding varieties of wheat. Janakiramaraju[31] and Mannivannan[43] reported that mass media exposure and contact with extension agencies had significant and positive association with knowledge level of farmers.

2.3.2 Adoption of Improved Practices

Nagendranath[48] reported that young farmers were better adopters of dryland practices than aged farmers. Reddy and Reddy[70] observed positive and significant relationship between age and adoption of the dryland farming practices.

Many of the past studies, Desai,[31] Mannivannan[43] and Vijayaraghavan[101] have indicated the presence of a positive relationship between educational level of farmers and their adoption behaviour. However, Veerbhadraiah and Dwarikanath[97] reported that there was no significant association between the educational level of farmers and their adoption behaviour. Nagendranath[48] reported that higher the education of beneficiary, greater was the scope of adoption of certain improved agricultural practices. Reddy and Reddy[70] reported positive and significant relationship of education with the adoption of dryland farmers.

Reddy and Reddy[70] reported the significant contribution of the variable, family size in the prediction of castor adoption of dryland farmers. Balkrishan Singh *et al.*[4] also observed the positive and significant correlation of family size with the farmers' adoption of dairy innovations and even this attribute had significant influence on the adoption of dairy innovations of the farmers, both in progressive and non-progressive villages.

Most of the studies have revealed that there was positive association between the size of land holding and adoption behaviour[13,43,55]. A contrasting tendency was observed by Jaiswal[28], who reported that size of farm had no relation to adoption of practices.

Many studies indicated that the social participation and adoption of recommended practices were positively related[13,48,70]. However, Mannivannan[43] and Veerbhadraiah and Dwarkanath[87] indicated that there was no association between the social participation and adoption behaviour of farmers.

Nagendranath[48] reported that small farmers having better socio- economic status were better adopter of certain dryland practices than marginal farmers. Most of the researchers observed a significant relationship between socio-economic status and adoption of farmers[16,94]. Gartrell and Gartrell[21] pointed out that the social status is the main cause for variation in the adoption behaviour of the farmers under Indian conditions. Doijad and Sen[17] stated that socio-economic status of the farmers had significant relationship in the adoption of improved practices.

Farmers having more cropping intensity possess more knowledge which leads to more adoption. Kahlon[36] noted that multiple cropping increase the alternative and scopes for farm income. It leads to higher adoption leading to increased production. Same thought is reflected by Gomez[22] in his study. Reddy and Reddy[70] reported that cropping intensity was significantly and positively correlated with Jowar adoption.

Solane[87] emphasized that every effort must be made to

remove Institutional bottlenecks involved in providing adequate and timely supply of modern inputs and credit which ultimately influence adoption.

Somasundaram[88] reported that economic motivation had significant relationship with the adoption behaviour of small farmers. Janakiramaraju[31] reported that economic motivation did not show any significant relationship with the extent of adoption of fertilizer practices amongst farmers of irrigated and non-irrigated areas.

Singh[80] reported that risk orientation was positively related to the adoption behaviour of farmers. Ganasekaran[20] found that risk preference did not show any relationship with adoption in both paddy and millet growing marginal farmers. Reddy and Reddy[70] observed significant and positive relationship of risk preference with the adoption of dryland practices.

Supe and Salode[95] reported that scientifically oriented farmers had high adoption levels. Dhanorkar[15] stated that scientific attitude towards farming helped the farmers in thinking rationally, with the result their knowledge and adoption level increased.

Several researchers have emphasised the role of mass media as well as the personal cosmopolite sources of information in influencing the adoption behaviour of the farmers. Rajguru and Satpathy[60] stated that although mass media like radio and newspapers were found to be effective in helping the farmers, in rational decision making they could not influence the adoption of innovation. In case of finally accepting the practices, personal cosmopolite sources (extension workers and dealers) were found to be more effective. Sen and Doijad[73] stated that mass media and personal cosmopolite sources of information had positive and significant relationship in the adoption of pest management practices by the farmers. The extent of utilisation of personal localite sources of information by the farmers were however found to be negatively associated with their adoption of plant protection practices.

Knowledge as a factor affecting adoption behaviour has been studied by the most of the researchers[26,41,46,69,82]. In their studies, they reported positive association between knowledge and adoption of improved agricultural practices.

2.3.3 Attitude of Farmers

Past researches conducted by Jaiswa[26], Moulik[46] and Singh[80] showed positive and significant relationship between extent of adoption of agricultural innovations and the attitude of the farmers. Sen and Doijad[74] supported the above views while studying the relationship of attitude of the farmers with the adoption of integrated pest management practices.

Sudarshan Rao *et al.*,[91] found that variables like extension contact, social participation, mass media participation, scientific orientation of farmers were significantly and positively correlated with their attitude. Further, economic orientation and participation in the extension activities of the respondent farmers were contributing to the prediction of the attitude of farmers. Reddy and Reddy[70] found that cropping intensity and risk orientation were positively and significantly correlated with attitude. Mathiazhagan and Singh[45] observed that education and farm size were positively and significantly correlated with attitude towards banana cultivation.

2.3.4 Skill

Skill implies the routine and periodical maintenance of the appliance including the minor repair and replacement of worn out parts. If the appliance is not maintained in a good condition then it may not deliver the expected work. The sprayers, dusters and fumigators used in the plant protection operations need to be maintained properly so as to avoid hazardous effects or accidents. All these things need more mechanical skill on the part of the operator to handle these appliances effectively. In other words, it is the ability to achieve the highest efficiency from the implements or machines in operation.

2.4 PROBLEMS OF TRANSFER OF TECHNOLOGY

Reed *et al.*,[68] reported that the lack of integrated approach created havoc among the farmers of doubling dryland agriculture with respect to the polyphagus pest namely *Heliothus armigera*. The greatest potential and challenge for the future would appear to lie in a real integrated approach to manage the pests in such areas. Laxminarayana[40] stated that a strategy for pest management needs to be tailored to tackle the problems in a particular region. Further, he also emphasized the need towards undertaking the research on biological control of pests. Pradhan[56] pointed out the serious lacuna in pest control research which acted as a drain across the road to progress of Indian agriculture. He further stressed that the different protection sciences can not be combined in any one individual subject and rethinking must be made on this issue. In plant protection uncertainities play vital role. Jock[34] reported that the uncertainities like rains and temperature fluctuations were the main constraints in the technology transfer of plant protection among the dryland farmers. He also further stated that the uncertainities will be the new challenges for future researchers.

Bhatnagar[7] revealed that though technologies were available even in the vicinity but they were not applicable and useful due to their sophistication. They were not translated into simple and understandable form. Bulk of technologies were unsuitable to the agro-climatic and socio-economic conditions of the people which restricted its further transfer to the needy people. Rarely, the absorption status (i.e. needs, resources etc.) were considered by the researchers while developing and recommending the technologies to the farmers.

Lack of confidence in extension agency is one of the problems of transfer of technology among farmers. Raje[58] pointed out that the activities of the Department of Agriculture are input supply oriented rather than education oriented. Unhealthy attitude of farmers towards development projects due to subsidy to the farmers has resulted in lack of confidence on the extension agency. Dak and Purohit[12] pointed out that the available information on new dryland farming technology released by the Department of Agriculture, Haryana in consultation with Haryana Agricultural University, Hissar did not percolate down to the change agents. Within the area of agronomic practices, a greater information gap among change agents was noted in fertilizer practices followed by plant protection practices and sowing practices. Lack of effective extension system in certain areas, deficiencies in the organisational structure and the target oriented approach to development are some of the hinderances in transfer of technology.

Knowledge is an important input in transfer of technology. Dak and Purohit[12]pointed out the poorly conducted pre-seasonal state level training workshop is responsible for the training information gap among the change agents (extension workers.)

Raje[58] observed that workshop on production technology was not attended by grass root level extension workers and most of the information was lost in the long channel involved. Bhatnagar[7]stated that training is becoming synonym to tourism in few cases, because of arbitrary selection of training programmes, organisations, objectives, trainers, trainees and the subject matter. It is rarely need based. There are no linkage for transfer ability of training and are rare opportunities for trainers' training. Even in good training programme there is more emphasis on knowledge rather than development of technical, communicational and managerial skills.

There is lack of storage facilities of inputs at village level. The inputs need to be prepositioned in the villages, and need to be made available to the farmers at the time of need and before the commencement of crop season. Unfortunately, this has not been done in majority of the cases. Non-availability of plant protection appliances and its repairing facilities is another constraint in its use[7]. In addition to the above there are certain problems of transfer of technology at the farmers' level.

Reddy[69] pointed out that the majority of respondents reported the lack of knowledge about the chemical control of weeds as the main reason for its non-adoption. Singh[80] found that the reasons for non-adoption of improved farm practices by the farmers were chiefly, the lack of knowledge, resources, innovativeness, prospective (motivational) price and attractive market.

Rai[57] found that lack of technical know-how, timely supply of seeds and fertilizers, high initial investment and adverse agrophysical conditions were the major factors associated with the non-adoption of improved agricultural practices.

Jaiswal and Singh[27] observed that the major factors associated with low level of adoption of improved agricultural practices were lack of adequate irrigational facilities, poor economic condition of farmers, lack of timely and adequate supply of protection requisites, ineffective demonstrations and poor contact of extension workers with farmers.

Bhaskaram[5] reported that reasons frequently stated by the respondents for non-adoption of some practices were the lack of awareness of the practices, non-availability of inputs and their high cost, lack of technical guidance and unsatisfactory personal experiences. Veerabhadraiah and Dwarakinath[97] pointed out that the farmers did not adopt recommended quantities of fertilizers due to lack of knowledge about the practice and resources as a reason for non-adoption. Rajagopalan and Singh[51] reported that ignorance and tradition stood in the way of following plant protection measures. The required chemicals were costly and thus poverty also stood in the way.

Reddy[9] found that lack of capital, complete knowledge about the practice, experience and conviction about the practice were found to be the important reasons for non-adoption of innovations. Sundaraswamy[93] reported that lack of knowledge and money were the main reasons for non-adoption of recommended hybrid jowar cultivation practices. Rao[64] found that non-availability of inputs was the main problems faced by users of farm machinery. Ambalagan[1] found that the major limiting factors for the adoption of the practice for paddy were lack of knowledge and non-availability of inputs such as fertilisers, plant protection equipments and high cost of cultivation.

Hans *et al.*,[25] stated that Indian dryland farmers perceive high levels of income risk. In India, the use of improved seed, fertilizers and plant protection measures is negligible. The available farm yard manure is spread thinly and farmers prefer to use it for cash crops and irrigated crops only. Low resource base and risk aversion are the two principle reasons responsible for the low adoption of technology. Rastogi[67] pointed out that hardly any plant protection measures are used in dryland areas. Plant protection chemicals application is extremely low. It is applied and used in a limited way to improved varieties and to in less than the recommended doses. Jodha[35] pointed out that the farmers' decision is largely guided by economic costs involved from the stand point of the majority of dryland farmers. In India, the high-yielding variety technology can be a high cost technology as it consists of costly inputs such as fertilizers, pesticides and better management of the crop. The farmers are reluctant to divert costly inputs meant for high yielding varieties to other non- high yielding crops.

Reed *et al.*,[68] also stated that all too often, the farmers spray too late, after they have seen the damage. Ideally, spraying against *Heliothus* should start as soon as large number of eggs are seen. The measures as recommended are being adopted in India, only when the pests or diseases assume serious proportions. It is probable that most farmers use no pesticides, not because they are unaware of pest losses or of the benefits to be obtained from pesticide use but because they do not have ready access to a practical, economic package of pesticide application. Sinha *et al.*,[85] stated that non-availability of fungicides and pesticides at proper time and lack of knowledge were the main reasons for non-adoption of plant protection practices.

Kunnal *et al.*, [39] pointed out that the application of plant protection chemicals to the crops was miserably low. High cost of chemicals, lack of knowledge, difficulties in application, lack of technical guidance and lack of conviction were some of the reasons for non-adoption.

Government of Bihar[24] pointed out that expensive input, lack of irrigation, unavailability of input and lack of advice were the reasons for non-adoption or partial adoption of plant protection practices by the farmers.

Lack of know-how, lack of skill in handling and repairing the equipments, high initial cost of the equipments were some of the constraints observed by Rao[65] in the use of improved plant protection equipment.

Report of the Evaluation Committee on Lab to Land programme, Indian Council of Agricultural Research, New Delhi[2] pointed out that plant protection was the major constraint for crop production. High cost of chemicals, non-availability of plant protection chemicals and equipment at critical period in the villages were the problems in effective transfer of plant protection technology at the farmers' level as pointed out by the committee.

2.5. CONCEPTUAL MODEL

2.5.1 Transfer of Technology

The phrase 'Transfer of technology' has a chequerred history. It began to be used on the International stage in discussion of the role of large and particularly multinational firms in the transfer of advanced production methods to developing countries entering the path of industrialisation. It has more recently come to be used in agricultural development. Transfer of Technology is a process by which the recommended practices produced by research and development are transmitted through extension agent to the farmers. In the term 'Technology', the appropriate technology is assumed[8]. Further Bunting[8] stated that a large number of appropriate farm technologies have been developed at the research stations, but only a few of these have been adopted at the farmers' level. The factors influencing the transfer of technology from research institutes to the users need to be critically examined particularly for promoting adoption of farm innovations by the farmers.

In the process of recent agricultural development, Dwarikanath and Channegowda[18] observed that the prime mover is considered to be the new agricultural technology. The benefit of such a technology is actually derived only when it is utilised efficiently by individual farmers in their local situation. If this is to happen, the new technology should be transmitted from source where it is generated to the farms where it is to be applied. It needs to be so represented as to become acceptable to the adopters. Dwarikanath and Channegowda stated that three systems, namely, research, extension and farming or clientle system are involved in the transfer of technology at the farmers' level. Jaiswal and Arya[29] added fourth system i.e. support system to the three systems stated by Dwarikanath and Channegowda[18]. The research system takes care of the technology production. Extension on system, dissemination or communication system, transfer the technology from research station to the farmers field. Support system helps the technology transfer process and provide necessary inputs required for the use of technology. Farmers' system is the technology utilisation system which adopts the technology and takes benefit from it.

Information regarding improved technology released by the research system can reach the farmers either through the extension system or directly to the farmers through mass media like radio, television and printed media, personal cosmopolite and localite sources of information. After receiving the information, in the innovation decision process, the farmer may either mentally accept the practice or reject it. In case the practice is accepted by the farmers, infrastructure support for timely and adequate supply of inputs is needed to facilitate actual use of the practice by the farmers. The non-availability of inputs naturally lead to non-adoption of the practice. Adoption of innovation involves a decision by the farmers for its continued use and integration into their farming practices[71].

2.5.2. The Problem-Complex to Transfer of Technology

Not only a single individual system is responsible for effective transfer of technology at farmers' level, but also the diverse sets of factors act as 'problem complex,'* in the transfer of technology process as evidenced by the number of social science researchers. All the factors separately diagnoised as being responsible for varying degree of adoption of improved agricultural technology do not act independently or in isolation. They interact with each other to from research a problem complex This complex of resistance, with manifold intensity and strength, acts as an impediment to transfer of technology from research station to the farmers. The pattern of the problem complex varies from one innovation to another and one farming community to another. It is for this reason that any programme aimed at transfer of technology must take into account the macro-level and micro-level factors related to the farmers, along with the quality of the extension approach and the characteristic of the improved technology.

2.5.3. Conceptual model for the Present Investigation

Kallen[37] stated that, "Every so often, when I come upon the word 'problem' in a work of philosophy or social science, I experience wonder. Their problems seem to follow from their solutions, not their solutions from their problems. Innately, we prefer bclief to doubt and certainity to uncertainity, we feel safer with ordered sequence than with interminate change, 'repetitive connection' gives us assurance, innovative variation does not".

In designing a study on problems of transfer of Technology of Plant Protection among the dryland farmers', it was opt to develop an appropriate conceptual model in which cognizance of the above thought provoking view of

---

* Problem is a word synonymous to other words like 'force', 'check', 'bind', 'curb', etc. the word 'constraint', is alternatively used for problem, which means 'barrier', 'bottleneck', 'hindrance', impediment', 'road block ', 'set back', etc.

Kallen was taken. The conceptual model in the present investigation depicts the relationship between the independent and dependent variables in the study. It is the ultimate result of the objectives to be studied, review of literature and indepth discussion with experienced scientists and extension workers in the field of transfer of technology of plant protection particularly for the dryland farmers. Such a model will not only guide the research under report but will also set a tone for further investigations that might be taken in this area.

Plant protection status allows us to know as to how much knowledge the farmer possesses about scientific plant protection practices, his attitude towards plant protection and his skill in controlling the pests along with his adoption[51]. In the conceptual model of the present study it is postulated that plant protection status of a farmer is a special kind of situation which helps in acceptance of plant protection practices. This plant protection status is a function of many variables. The important variables affecting plant protection status are age, education, family size, land holding, social participation, socio-economic status, availability of plant protection equipments, agricultural, infrastructure, cropping intensity', economic motivation, scientific orientation, risk preference, utilisation of sources of information and adoption. The relationship of these variables with plant protection status is to be found out in order to see the possibility of manipulating these variables for increasing the pest control efforts of the dryland farmers. A similar set of variables was also tested by Pande[51], and it was observed that education, income, availability of plant protection equipments, scientific orientation, economic motivation, risk preference and socio-economic status were positively and significantly related with the plant protection status of the cotton growing farmers. All the above 14 variables undertaken for the present study were categorised as socio-personal, socio-economical, situational, psychological and communicational variables which comprise the comprehensive set of independent variables which ultimately affect the dependent variables i.e. plant protection status of the dryland farmers. The same set of variables

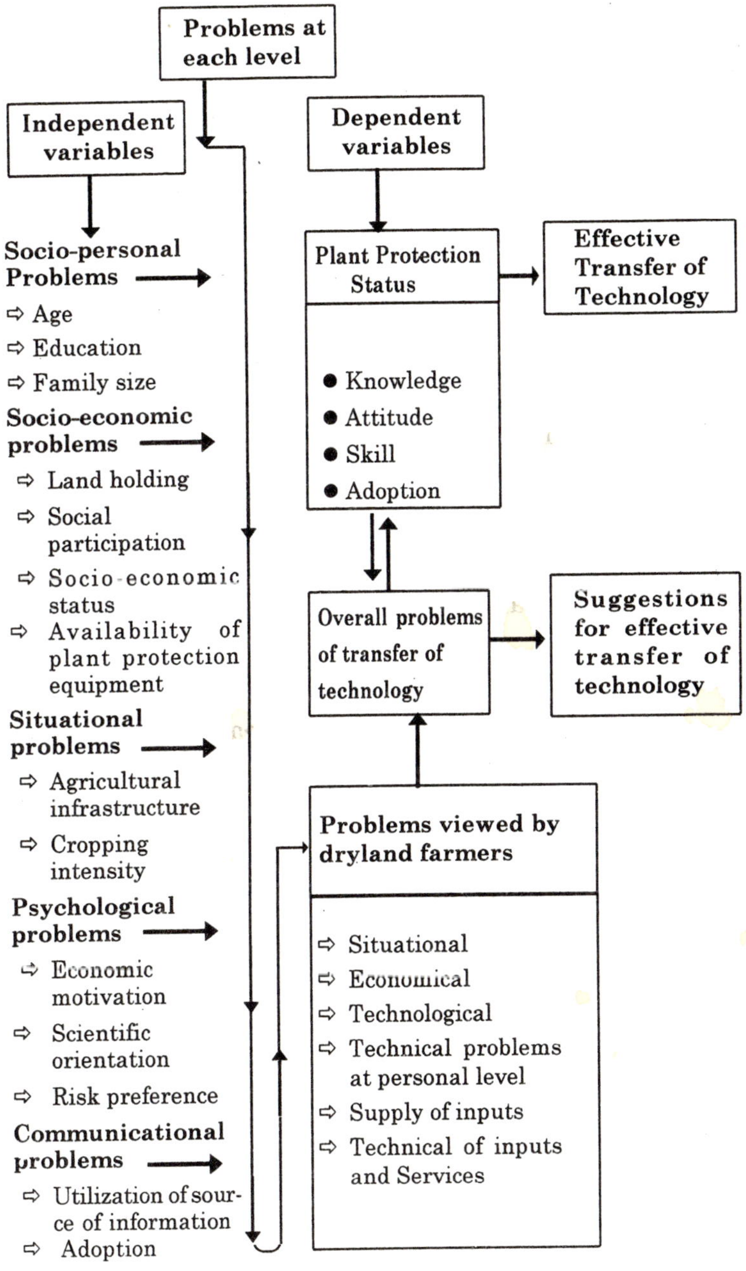

Fig. 1: Conceptual model on problems of transfer of plant protection technology among dryland farmers.

under each category also act as problems and as a whole a 'problem complex' or factors in affecting the plant protection status of the dryland farmers which ultimately affect the transfer of plant protection technology at the farmers' level adoption.

Besides these, problems as viewed by the dryland farmers were also considered in the present investigation. These problems were categorised as situational, economical, technological, technical problems at personal level, problems at supply level and problems of technical guidance and services. Thus, as a whole, it depicted an overall problem complex which affect in the variation of plant protection of an individual farmers, ultimately affecting the transfer of plant protection technology at their level of adoption. To minimise the 'overall problem-complex' the implications, emerging from the present investigation, would be helpful for effective transfer of technology at the farmers' levels in adopting the plant protection technology.

2.6. DEFINITION OF THE TERMS USED

Age : Number of chronological years completed by the respondent on the previous birth day.

Education: Formal education level of the respondent[99].

Family size : The number of family members in the family[99].

Land holding : It denotes an area of land possesses by an individual[99].

Social participation: Operationally, it had been defined as the degree to which the respondents are involved in formal social and political organisations as members or office bearers or involvement in community work[99].

Socio-economic status : It is the position an individual occupies with reference to the prevaling average standard or cultural positions, material possession and participation in group activities of the community[99].

Availability of plant protection equipment : It is material possession of plant protection equipment with an individual respondent.

Agricultural infrastructure : This variable is defined as the physical capital institution and organisation, both public and private which provide economic service and which have a significant effect upon the economic functioning of the individual farm/firm, but which are external to the separate individual farm/firm[72].

Cropping intensity: It is the indicator of the extent of utilisation of land resources, dividing the gross cultivated area by the net cultivated area and thus the cropping intensity is expressed in percentage.

Economic motivation: It is the occupational success measured in terms of profit maximisation and the relative value placed by an individual on economic ends[94].

Scientific orientation: It is the degree to which an individual is inclined to use scientific methods in farming and decision-making[94].

Risk preference : It is the degree to which farmers are oriented towards risk and have courage toface the new problems of uncertainity in farming[94].

Utilisation of sources of information: It is the degree of utilisation of different sources of information by the farmers; such as mass media, personal cosmopolite and personal localite sources of information to obtain information about plant protection.

Source credibility : It refers to the relative importance attached to the sources of information used by the farmers in adoption of plant protection.

Adoption: It is the acceptance and use of plant protection practices on a continued basis on chilli crop.

Plant protection status : It refers to the knowledge,skill and adoption of farmers towards plant protection practices[51].

3

RESEARCH METHODOLOGY

Chapter three contains the research methodology in which sampling, selection of variables, measurement of variables and to study the problems of transfer of plant protection technology have been dealt.

3.1. SAMPLING

3.1.1 Background Information of Study Area

The study was conducted in Vidarbha region of Maharashtra state in which Dryland Farming Research Centre of Akola cater the needs of dryland farmers residing in this area. It is semi-arid region having mean rainfall 830 mm from June to September of which 50 per cent is received during July-August Soils are black, medium and deep sandy loams to heavy clays clacarus, with lime kankars at varying depths and are highly deficient in available phosphorus. They have high phosphate fixing capacity and also have high moisture retentivity but develop deep cracks from December onwards. Farmers predominantly take kharif crops but double cropping is possible in some of the areas where rainfall receives more than mean annual rainfall, i.e. 830 mm. The major crops grown are cotton, chillies; cash; crops sorghum, paddy-food crops safflower and lentil as oilseed crops.

3.1.2 Selection of Crop for Study

In order to study the problems of transfer of plant protection technology, it was essential to select one particular crop in which dryland farmers adopt plant protection measures. It is evident from the literature that dryland farmers do not take any plant protection measure for the crops except cash crops. Cotton and chillies are the main cash crops grown in the Vidarbha region on which mostly plant protection measures are taken by the dryland farmers as observed in the pilot study conducted for the present investigation. Chilli crop has been purposively selected so as to study the objectives decided.

Chilli is an important cash crop in India. It comes under the category of spices and condiments. It is grown for its pungency to the food. As a condiment, it has become indispensable in every Indian home in curries and pickles. It is also used in medicine. It is grown in all parts of India. The area under chilli was 877.8 thousand hectares with 708 thousand tonnes of production in 1985-86 in India.

The total area under chilli cultivation in Maharashtra in the year 1985-86 was 143.7 thousand hectares with the total production of 71.5 thousand tonnes. In Vidarbha chilli is grown in almost all the districts in 57.11 thousand hectares with a production of 21.80 thousand tonnes.

Chilli is the important cash crop in Vidarbha region. Since last few years, the area under chilli was found to be declining. The important reason for this decline is the infestation of pest and disease attack on this crop. There is growing tendency among farmers not to adopt plant protection measures for this cash crop. The results of the present study may be useful directly or indirectly to the chilli growing farmers as well as planners, executives, extension workers in effective transfer of plant protection technology among dryland farmers.

3.1.3 Location of the Study Area

The study was conducted in purposively selected dryland block, Kuhi in Nagpur district of Vidarbha region where most of the farmers grow chilli crop. Panchayat

Samiti Kuhi is situated adjacent to Nagpur block with Kuhi as a block headquarter. It is 25 miles away from Nagpur. Kuhi and Mandhal are the big villages in this block, having the population above ten thousands. Kuhi is situated near Nagpur-Nagbhir railway line. Railway line passes through the block from North-South at the one corner of the block. But railway is not suitable for communication of the villagers in the block. State transport buses are used for communication. There are eighty Gram panchayats in this Panchayat Samiti. Geographical area is 87.645 hectares out of which 2,500 hectares are under forest land, 23,100 hectares are fallow and 62,045 hectares are under cultivation. The soil is medium black. In kharif season paddy, chillies and tur are grown. Paddy and chillies are the main, crops of this area. In rabi season wheat, jowar, gram, linseed and ~~til~~ are generally grown. Vegetables and fruit gardening are not practised by the cultivators. The average rainfall is 1,000 mm and the rainy season is in between the June and September months. Minimum and maximum temperature is 53°F and 116°F. Climate is suitable for both crops and the life.

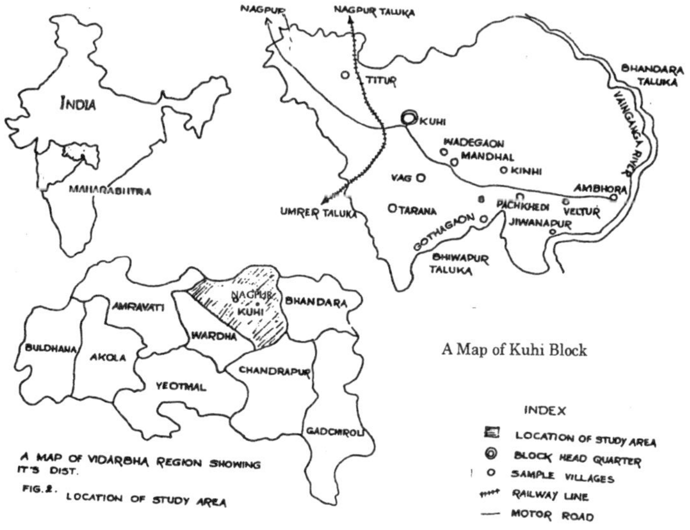

Fig. 2 : Location of Study Area

3.1.4 Selection of villages and farmers

The villages selected for the study were maximum chilli growing villages. Out of the total villages, 13 villages were selected for study from which the sample was drawn, From each village, a list of chilli growing farmers was prepared with the help of Patwari record. From this list of each village, one tenth sample was selected by equal interval method of randomisation. Thus, a sample of 200 farmers was drawn from whom the data for the study was gathered.

3.1.5. Pilot study

A pilot study was conducted before finalising the schedule for investigation. The site for the try-out study was selected by taking into consideration the nature of problem. This being exploratory study in some aspects; it was a problem to decide items for schedule preparation. Hence, a pilot study was undertaken as a forerunner of main schedule preparation.

3.2 SELECTION OF VARIABLES

3.2.1 Independent

Socio-personal	:	Age, education, family size
Socio-economic	:	Land holding, social participation, socio-economic status, availability of plant protection equipment.
Situational	:	Agricultural infrastructure, cropping intensity.
Psychological	:	Economic motivation, scientific orientation, risk preference.
Communication	:	Utilisation of sources of information, adoption.

3.2.2 Dependent

Plant protection status

Table 1 : Variables studied and their measurement		
Sr. No.	*Variables*	*Measurement*
A.	INDEPENDENT	
	Socio-personal	
1.	Age	
2.	Education	Scale developed by
3.	Family size	Venkatramaiah [99]
	Socio-economical	
4.	Land holding	Scale developed by
5.	Social participation	Venkatramaiah [99].
6.	Socio-economic status	
7.	Availability of PP equipments	Structured schedule
8.	Agril. infrastructure	Structured schedule
9.	Cropping intensity	Structured schedule
	Psychological	
10.	Economic motivation	Scale developed by
11.	Scientific orientation	Supe[94]
12.	Risk preference	
	Communicational	
13.	Utilisation of sources of information	Structured schedule
14.	Adoption	Structured schedule
B.	DEPENDENT	
15.	Plant protection status	Plant protection status index developed by Pande[51]

3.3 MEASUREMENT OF VARIABLES

At a glance, different variables and their measurement are presented in Table 1.

3.3.1 Independent Variables

3.3.1.1 Socio-personal

Age : Number of chronological years completed by an

individual on the last birth day has been considered as age of that particular respondent and categorised as young (25-35 years), middle (36-45 years) and upper age (45 and above).

Education : To quantify the education of the respondents, the scoring system followed by Venkatramaiah[135]in his socio-economic scale was used. The educational status was noted against the items composed of, no school education, functional literal, primary school, middle school, high school and college with corresponding scores 0,1, 2, 3, and 5, respectively and categorised under the present investigation as illiterate, up to primary, middle school, high school and above.

Family size : Here also the scoring system followed by Venkatramaiah[135] in his socio-economic status scale was utilised and categorised as small (1-3 members), medium 4-6 members), large (7-9 members), very large (10 and above) and special features with corresponding scores 1, 2, 3, 4 and 5 respectively. Under present investigation it was categorised as small family (1-6 members) and big family (more than 7 members).

3.3.1.2 Socio-economic

Land holding : Different researchers have tried to measure the land holding in different ways. Venkatramaiah[135] while constructing socio-economic status scale, used item analysis and determined weightage for different size of land holding group. In this scale the scores were : farmers with no land - 0, up to 1 hectare -1, up to 2 hectares- 2, up to 4 hectares -3; up to 10 hectares- 4 and above 10 hectares -5 and categorised under the present investigation as small and marginal (1-2 hectares , semi-medium (2.1-4 hectares); medium (4.1-10 hectares); and large (above 10 hectares).

Social participation : To quantify the social participation, the scoring of items was done according to the socio-economic status scale developed by Venkatramaiah[135] (detail in Appendix-I) and categorised as low participation (0-2 scores) and high social participation (above 3).

Socio-economic status : Socio-economic status scale developed by Venkatramaiah[135] was used to measure the socio-economic status and its (items) components. Scoring of each item is given in Appendix-I. It was categorised as :

Low (below 15)	:	Less than mean- standard deviation
Medium (16-29)	:	Between mean-standard deviation to mean + standard deviation
High (above 30)	:	Above mean + standard deviation

Availability of plant protection equipment : In order to measure the availability of plant protection equipment, farmers were asked to give their responses about plant protection equipment on two points, i.e. available and not-available with corresponding scores '1' and '0' respectively and termed as availability of plant protection equipment score.

3.3.1.3 Situational variables

Agricultural infrastructure : To measure the agricultural infrastructure, a structured schedule was utilized. The schedule composed of ten areas of agricultural infrastructure facilities like (1) irrigation (2) credit institution, (3) electricity, (4) input supply agencies, (5) transportation, (6) communication, (7) media for agricultural study, (8) marketing, (9) land policy, and (10) storage. The responses against each of these statements were noted against a five point scale comprising of very high facility, high facility, average facility, facility at unsatisfactory level and facility at a very insignificant level with corresponding scores 4, 3, 2, 1, and 0 respectively. The total of all the scores obtained against each statement termed as agricultural infrastructure score and categorised as:

Low (below 11)	:	Less than mean-standard deviation
Medium (12-25)	:	Between mean-standard deviation to mean standard deviation
High (above 25)	:	Above mean + standard deviation

Cropping intensity (utilisation of land resources): Our general expectation for agricultural production is that farmers who utilise their land more intensively usually get more production by way of their increase in knowledge, skill, attitude and adoption. This view had been supported by many researchers.

Kahlon[36] in his study noted that multiple cropping increases the alternatives and scopes for farm income as it leads towards increased production. Some thoughts were reflected in the study by Gomez[22] where it was observed that increased agricultural production can be secured only by intensification of use of the land resources .

Depending on the review of the available literature and results of past studies, cropping intensity of the respondents from was taken, as the indicator to the extent of utilization of land resources. In order to calculate the cropping intensity, the gross area under crops during the whole year divided by the respondents net cultivated area and was expressed in terms of percentage and categorised as :

Low (below 124)	: Less than mean-standard deviation
Medium (125-152)	: Between mean-standard deviation to mean + standard deviation.
High (above 153)	: Above mean + standard deviation

3.3.1.4. Psychological

Economic motivation, scientific orientation and risk preference:

Three value scales developed by Supe[94] were used to measure economic motivation, scientific orientation and risk preference. Each scale had 6 statements and these statements were scored on five point continuum. Each scale had some positive and some negative statements. In economic motivation scale, item number 1, 2, 3, 4 and 5 were positive statements, while item number 6 was negative. In scientific orientation scale item number 1, 3, 4, 5 and 6 were positive and second statement was negative. In the risk preference scale, item number 2, 3, 4 and 6 were positive, while item number 1 and 5 were negative statements.

The five points on which the statements were scored were strongly agree, agree undecided, disagree and strongly disagree, The scoring was '7' for strongly agree, '5' for agree, '4' for undecided, '3' for disagree and '2' for strongly disagree, while for a negative statement scoring was just reverse. The total score for a particular scale was obtained from the scoring of six items included in that particular scale and termed as score for economic motivation, scientific orientation and risk preference and categorised as follows:

	Low	Medium	High
Economic motivation	below 27	28-39	above 40
Scientific orientation	below 25	25-41	42 and above
Risk preference	below 23	24-39	(above 40)

The above categorisation had been done by following criteria, which is as follows:

Low : Less than mean - standard deviation

Medium : Between mean- standard deviation to mean + standard deviation

High : Above mean + standard deviation

3.3.1.5 Communicational

Utilisation of sources of information : Considering the different past studies on the sources of information some of the sources of information were selected, which were found to be more frequently used by the farmers in getting the information regarding plant protection practices, particularly in relation with the chilli crop during the pilot study. The selected sources of information were classified into 3 groups, i.e. utilization of mass media, personal cosmopolite and personal localite sources of information with a view to compute a score for complete utilization of sources of information, the subcomponents mentioned in the schedule (Appendix-I) were taken into account.

The respondents were asked to indicate that how often they utilized these sources of information against a five point scale consisting of 'most often', often', 'sometimes', 'rarely' and 'never' with corresponding scores 4,3,2,1 and 0, respectively. The total of all the scores of utilization of mass media personal cosmopolite and personal localite sources of information is termed as a score for utilization of sources of information and categorised as:

Low (below16)	:	Less than mean - standard deviation
Medium (17-40)	:	Between mean-standard deviation to mean + standard deviation
High (above 41)	:	Above mean+ standard deviation

Sen[72] utilised the same technique as in the present investigation.

Adoption : A List of the recommendation of plant protection practices for chilli by Punjabrao Krishi Vidyapeeth, Akola and the Department of Agriculture was prepared. From the list of plant protection practices on chilli crop, three preventive and three curative practices were selected for studying the adoption in consultation with the Chief Chilli Scientist, Entomologist and Plant Pathologist of Punjabrao Krishi Vidyapeeth, Akola.

The six selected plant protection practices along with their details are given below :

(i) Seed treatment : To control dieback, chilli seeds should be treated before sowing, with Agrosan at the rate of 3 gms/kg of seed or Thirum at the rate of 2 gms/kg of seed.

(ii) Soil treatment : Soil application of 10 per cent phorate granules should be applied at the rate of 10 kg

hectare to avoid the attack of insect pests on chilli crop in future. These granules may be put in the prepared lines of seed bed and be covered with soil for avoiding its direct contact with seed.

(iii) Seedling root-shoot dip : To avoid insect pest attack before flowering, the whole seedlings of the chilli should be dipped, in the solution of 4 gm of DDT 50 per cent in 1 litre of water.

(iv) Spraying of insecticide plus pesticide after transplanting and before flowering : In order to check and control leaf curl (Churda-murda) and dieback the following solution of insecticide plus pesticide should be sprayed 10 days after transplanting,

Nuvacron (monocrotophos) 40 per cent systemic, 1-2 ml + Copper oxchloride 2.5 gms or Dithane z-78 or Dithane M-45 or Cuman L+ water, 1 litre.

150 to 200 litres water will be required by hand operated spray pump. Three subsequent sprayings of the above solution should be prepared and sprayed after 10 days interval and continued upto flowering stage. About 3 sprayings may be given before flowering.

(v) Spraying of insecticide plus pesticide after flowering : In order to control Churda murda and dieback, about 3 sprayings of the following mixture should be sprayed after flowering, at every 10 days interval. Spraying may be taken after harvesting of chillies.

Malathion 50 per cent systemic 2 ml

or Endosulphan 35 per cent systemic 1.5 ml

or Monocrotophos 40 per cent systemic 1-2 ml

+ Dithane z-78 or M-45 2.5 grams

+ 1 litre of water.

(vi) Spraying of water soluble sulphur : In order to avoid the attack of mites and its control water soluble sulphur should be sprayed at the rate of 2-3 grams per litre of water when the attack of mites start in the month of August and September.

In order to get the adoption score, farmers responses were taken for each of the above practices on extent of adoption, level of adoption of recommended doses and correct stage or time of application. In case of extent of adoption the responses were taken on full area, more than half, and less than half with corresponding scores, 3, 2 and 1, respectively. In case of level of adoption of the recommended doses, the responses on recommended dose below recommended and above recommended dose were taken with corresponding score 3, 2 and 1, respectively. For time of application the responses were taken on correct stage or time, before correct stage or time and after correct stage or time or time with corresponding score 3, 2 and 1, respectively. The total of all the scores for all the six practices obtained is termed as adoption score of a particularly respondent and categorised as :

Low (below 6)	:	Less than mean -standard deviation
Medium (7-42)	:	Between mean-standard deviation to mean + standard deviation
High (above 43)	:	Above mean + standard deviation

3.3.2 Dependent Variable

3.3.2.1 Plant protection status

For the purpose of measuring plant protection status of the dryland farmers plant protection status index developed by Pande[51] was used.

Components of plant protection status index have been categorised in four groups namely, knowledge, skill, attitude and adoption of the plant protection measures.

Knowledge is the situation characterised by the possession of technical know-how by an individual with regard to agronomic practices and plant protection of chilli crop. These two aspects of chilli production are very important and the latter further requires information about insecticides and their proper use, appliances, pest, their nature of damage and various types of control measures. Thus the individual is exposed to the innovations existance and gains some understanding of how it functions.

Skill is the state of attainment mastered by an individual engaged in agricultural profession acquiring thereby efficiency in identifying various pests, their stages, determining appropriate insecticides to be used in the control of pest with proper concentration and at proper time.

Attitude is a relatively enduring organization of an individual's beliefs about an object that predisposes his action. The attitude of an individual predisposes him to use pesticides and appliances more scientifically in the control of pests. It also provide him an ability to purchase proper insecticides and appliances along with other material needed for plant protection practices.

Adoption is a decision to make use of new idea as the best course of action available. This is actually a stage where action in terms of use of the plant protection practice commence. It confirms acceptance of different aspects of plant protection as one of the measures for enhancing yield of chilli per unit area.

Plant protection status index includes 34 knowledge statements, 10 attitude statements, 10 skill statements and 21 adoption statements under the four components of plant protection status index as shown in the Appendix-I. The item have been administered on a three point rating scale of strongly agree, agree and disagree with a scoring of 3, 2 and 1, respectively. The range of the scores was 75 to 225. The scores thus obtained is indicative of the plant protection status of the respondents and categorised as:

Low (below 111)	:	Less than mean-standard deviation
Medium (112-175)	:	Between mean- standard deviation to mean + standard deviation.
High (above 176)	:	Above mean + standard deviation

3.4. PROBLEMS OF TRANSFER OF TECHNOLOGY IN PLANT PROTECTION ADOPTION BY THE DRYLAND FARMERS

In order to study the various problems faced by the dryland farmers in the adoption of plant protection measures, the methodology followed in the present investigation is given below.

Based on the review of literature, discussion with the dryland farmers during the pilot study; a comprehensive list of the various problems faced by the dryland farmers in the adoption of plant protection measures was prepared. Depending on their nature, all these problems were classified into various categories namely, situational, economical, technological, technical problems at personal level, problems at supply level and problems of technical guidance and services. In each of the categories 1 to 10 numbers were given to the various problems. Thus, the entire list of the problems under each category was prepared on the separate sheet of paper and only number of these problems under each category were given in the final schedule (Appendix-I) . While interviewing the dryland farmers, the nature of the problem faced by them were ticked in the schedule for which numbers were allotted. In case, if the problem faced by them is not listed, all these problems were noted on the schedule itself. Thus, number of dryland farmers facing the particular problem were calculated in terms of percentages.

Various suggestions received from the dryland farmers for effective transfer of technology at the farmers' level were also noted in schedule.

3.5 DATA COLLECTION

3.5.1 Pre-testing of schedule

Interview schedule was prepared after studying the related background area connected with the investigation. The pilot study for getting relevant information, selection of block and villages was completed first. Then, the schedule was prepared by incorporating all the individual variables. Thus, the schedule so designed was subjected to pre-testing. The entire schedule was pretested from 10 farmers in the non-sample area of the project. On the basis of pre-testing, necessary modifications were made in the final schedule, (Appendix-I) incorporating appropriate measurements for selected variables and to make it more objective and understandable as far as possible to the farmers of the study area.

3.5.2 Collection of data

The schedule was also translated in local language 'Marathi' for an easy understanding to the respondents and for getting the correct required information while collecting data. Daily observations and important information not included in the schedule were also noted in the Diary to get in depth understanding about the area of investigation.

The data was collected during first February to the end of March, 1989, a period just after the peak period of picking red chilli. The assistance of local leaders, village extension workers, school teachers and progressive farmers were taken to establish rapport for getting the factual information. However, their presence during the actual interview was not solicited to avoid the bias of respondents. Thus, the farmers were individually interviewed when they were at leisure either at their farm or home by the researcher himself duly supported by personal observations. At the time of interview, the purpose of the study was explained to the respondents and only one respondent was interviewed at a time. The farmers' response were recorded on pre-structured schedule. Accordingly, the data was collected from 200 chilli growers from selected villages.

3.6 SCORING AND TABULATION OF DATA

After completion of data and relevant information the data were scored carefully. Before scoring, all entries of the schedule were further checked for completeness, consistency, legibility and uniformity in the nature of response. After this, the scores for each of the respondents in relation with each of the variables were written in the body of schedule. In order to facilitate tabulation; formats for tabulation were designed first and dummy tables with suitable captions were prepared itemwise.

3.7 STATISTICAL ANALYSIS

The data thus collected were scored compiled, tabulated and subjected to the statistical analysis. From all the scores obtained, percentages, ranking, means, standard deviation, coefficients of correlations, multiple regression and path analysis were carried out.

3.7.1 Statistical Tests Applied

Measures of Central Tendency and Variability : From all the scores obtained percentages, means, and standard deviations were calculated.

Correlation coefficients : The Pearson's product moment correlation coefficients (r) were computed between the independent and dependent variables in order to see their relationship with each other.

Multiple regression and path analysis : In order to understand the influence of the independent variables on plant protection status index, multiple regression analysis was carried out at the National Dairy Research Institute, Karnal computer.

The multiple linear regression equation model used in the study was:

$$Y = a + b_{1x2} + b_{2x2} \,.........\, b_{pxp} + e$$

where, Y is the dependent variable.

$X_1 X_2 ...... X_p$ are explanatory variables.

'a' is the constant term, and

'e' is the random error which follows normal distribution with mean zero and population standard deviation.

The partial regression coefficient b_1, b_2.....b_p were tested whether they were significantly different from zero by the use of 't' test with (n-p-1) degree of freedom. The 't' value was computed as follows:

$$t = \frac{b_i}{SE \text{ of } b_i}$$

The coefficient of multiple determination that is, R^2 was computed by the following formula :

$$R^2 = \frac{(bixiy)}{y^2 - \frac{(y)^2}{n}}$$

The significance of R^2 was ascertained by 'F' test by applying the formula as given below :

$$F = \frac{(n-p-1)\,R^2}{P\left(1-R^2\right)} \text{ with P and (n-p-1) d.f.}$$

The step down regression analysis was computed to determine the contribution of the significant variables eliminating the other non-significant variables. It was also observed that in the multiple regression analysis multi collineary problem was created due to high order of the inter correlation among the independent variables. To overcome this problem also step down regression analysis was followed. However, this analysis eliminated the insignificant variables but it failed to provide the desired signs for the partial regression coefficients. The alternative was dropping such influential variables. However, these variables could not be dropped as they are very highly significant in the regression equation and dropping these variables would lead to further reduction in R^2 (coefficient of multiple determination) value of the equation.

Further, path analysis was also computed in order to find out the direct and indirect effect, along with their channelising effect of the significant independent variables affecting the plant protection status of the dryland farmers.

4

RESULTS AND DISCUSSION

The results of empirical evidence collected in accordance with the objectives framed for this investigation have been organised and discussion thereon presented in this chapter. At the end empirical model has also been given along with its discussion.

4.1. CHARACTERISTICS OF DRYLAND FARMERS

Frequency distribution of respondents according to their different characteristics are presented in Table 2 and discussed as follows.

4.1.1 Socio-personal, socio-economic, situational psychological and communicational characteristics

The Table 2 shows that most of the respondents fell in the middle and upper age category, that is 34 and 46 per cent respectively and only 20 per cent respondents in young age category. As a whole, respondents were less educated. About one-fourth (24%) of the respondents were illiterate, followed by 30 per cent educated upto primary, 28 per cent upto middle standard and 18 per cent above high school level. Exactly equal, half-half of the whole respondents had small and big family size having 1-6 members in small family and more than 7 members in big family.

TABLE 2 : Distribution of the respondents according to their characteristics

Sr. No.	*Characteristics*	*Category*	*No. of respondents (N-200)*	*Percentage*
1.	Age	Young	40	20.00
		Middle Age	68	34.00
		Upper Age	92	46.00
2.	Education	Illiterate	48	24.00
		Upto Primary	60	30.00
		Middle	56	28.00
		High School and above	36	18.00
3.	Family Size	Small	100	50.00
		Big	100	50.00
4.	Land holding	Small and Marginal	48	24.00
		Semi-medium	32	16.00
		Medium	84	42.00
		High	36	18.00
5.	Social Participation	Low	128	64.00
		High	72	36.00
6.	Socio-economic Status	Low	40	20.00
		Medium	100	50.00
		High	60	30.00
7.	Availability of Plant Protection equipment	Available	92	46.00
		Not-available	108	54.00
8.	Agricultural infrastructure	Low	40	20.00
		Medium	112	56.00
		High	48	24.00
9.	Cropping intensity	Low	100	50.00
		Medium	72	36.00
		High	28	14.00
10.	Economic motivation	Low	40	20.00
		Medium	120	60.00
		High	40	20.00

TABLE 2 *(contd.)*

11.	Scientific orientation	Low Medium High	32 136 32	16.00 68.00 16.00
12.	Risk preference	Low Medium High	36 120 44	18.00 60.00 22.00
13.	Utilisation of sources of information	Low Medium High	52 108 40	26.00 54.00 20.00
14.	Adoption	Low Medium High	48 108 44	24.00 54.00 22.00
15.	Plant protection status	Low Medium High	36 120 44	18.00 60.00 22.00

Most of the respondents, that is 42 per cent had medium size of land holding followed by 24 per cent small and marginal farmers and only 18 per cent had high land holding. It was also observed that 64 per cent respondents had low whereas 36 per cent had high social participation.

Exactly half of the respondents (50%) had medium socio-economic status as against 20 per cent low and 30 per cent high. The data revealed that 46 per cent respondents had their own plant protection equipments whereas 54 per cent had not their own plant protection equipments such as sprayers and/or dusters. Respondents who had not their own plant protection equipment borrowed them from other farmers on hire basis.

More than 50 per cent respondents had medium agricultural infrastructure facilities followed by 20 per cent low and 24 per cent high. Exactly, 50 per cent respondents had low cropping intensity followed by 36 per cent medium and 14 per cent high. It is also a point to note that out of 100 respondents who fell in the category of low cropping intensity, 22 per cent of them had taken only a single crop during the year. Lack of irrigation facilities might be the root-cause for having the low cropping intensity.

With regard to psychological characteristics more than 60 per cent respondents had medium economic motivation, scientific orientation and risk preference.

The data revealed that 54 per cent respondents fell in the medium category of utilisation of sources of information followed by 26 per cent low and 20 per cent high category. In case of adoption of plant protection measures. More than half of the respondents (54%) fell in medium adoption while 24 per cent low and 22 per cent in high adoption category.

It was also found that 60 per cent respondents had medium, 22 per cent high and 18 per cent had low plant protection status.

4.1.2. Type of Plant Protection Equipments Available

The Table 3 consists of distribution of respondents according to the type of plant protection equipments with them. It is evident from the Table that out of 92 farmers who had plant protection equipments; most of them had only one sprayer, either knapsack or foot sprayer, that is 30 and 26 per cent respondents respectively followed by hand rotary duster (9%). Further, the respondents having medium size of land holding under chilli cultivation had one sprayer and one duster either knapsack sprayer and hand rotary duster or foot sprayer and duster. The Table 3 also indicates that 17 per cent respondents had foot sprayer and duster while 9 per cent had knap sack sprayer and duster. It also indicates that 9 per cent respondents had all the four type of plant protection equipments, i.e. knapsack sprayer, foot sprayer, hand rotary duster and power sprayer. Those who had all the above plant protection equipment were mostly rich farmers having big land holding under chilli, mostly the profit seekers who treated chilli cultivation as money earning profession.

Table 3 : Type of plant protection equipments available with the respondents

Sr. No.	*Name of plant protection equipments*	*No. of respondents (N-92)*	*Percentage*
1.	Knapsack sprayer	28	30
2.	Foot sprayer	24	26
3.	Duster	8	9
4.	Knapsack sprayer + duster	8	9
5.	Foot sprayer + duster	16	17
6.	Knapsack sprayer + foot sprayer + duster + power sprayer	8	9
	Total	**92**	**100**

Table 4 : Credibility of sources of information in the adoption of plant protection measures as perceived by the respondents

Sr. No.	*Source of information*	*No. of respondents (N-200)*	*Perce-ntage*	*Ranking (First choice by the respondents*
1.	Progressive farmers of the village	60	30	I
2.	Village Level Worker	48	24	II
3.	Agri. Exten. Officer	24	12	III
4.	Neighbours	20	10	IV
5.	Plant Protection dealers	12	6	V
6.	Friends	12	6	V
7.	Newspaper release	8	4	VI
8.	Publication of Agril. Dept.	8	4	VI
9.	Radio broadcasting	4	2	VII
10.	Progressive farmers of other villages	4	2	VII

4.1.3 Source Credibility

It is clear from the Table 4 that the progressive farmers of the village were considered to be the most credible source by the respondents amongst 15 sources of information utilised by them in adoption of plant protection measures. Thirty per cent respondents ranked first to the progressive farmers of the village as a credible source of information followed by Village Level Workers and Agri. Extn. Officers (24% and 12% respectively), neighbours (10%), plant protection dealers and friends (6 % each), newspaper release and publications of agricultural department (4 % each), radio broadcasting and progressive farmers of other villages (2 % each). The results obtained logical due to the fact that at the awareness stage, the respondents might have collected information through all 15 different sources of information, but particularly at the time of adoption of plant protection practices, the dryland farmers preferred more the progressive farmers of the village, Village Level Workers, Agri. Extn. Officers, neighbours, plant protection dealers and friends as credible sources of information in descending order. Being plant protection technology, a highly sophisticated one, dryland farmers might have put their first choice as a source of credibility mostly to the personal localite and personal cosmopolite sources of information so as to contact them from time to time, whenever required, to avoid any type of risk in all respects. From this, it could be interpreted that the progressive farmers in the villages might have acted as opinion leaders for the other farmers with regard to utilization of plant protection information. Further, it also observed that dryland farmers used more of competent and technologically sound sources of information, such as Village Level Workers and Agri. Extn. Officers, followed by non-institutionalised sources of information such as neighbours, plant protection dealers and friends which were of personal nature.

The findings are in conformity with the findings of Singh and Singh[78], Ramesh Babu and Sinha[61], Sinha *et al.*[86]

Table 5 : Farmers' awareness about recommended plant protection measures

Sr. No.	*Recommended plant protection measures*	*Awarness of farmers*		
		Number	*Percentage*	*Rank*
1.	Spray of insecticide plus pesticide before flowering	200	100.00	I
2.	Spray of insecticide plus pesticide after flowering	192	96.00	II
3.	Seedling root-shoot dip	180	90.00	III
4.	Seed treatment	148	74.00	IV
5.	Soil treatment	132	66.00	V
6.	Spray of sulphur	84	42.00	VI

*N = 200

4.2. ADOPTION AND RE-INVENTION OF THE RECOMMENDED PLANT PROTECTION MEASURES

4.2.1 Awareness

The Table 5 shows the extent of awareness of the six recommended plant protection measures. It is evident from this table that cent per cent farmers are aware of the spraying of insecticide plus pesticide before flowering, followed by 96 per cent of spraying of insecticide plus pesticide after flowering, 90 per cent of seedling root shoot dip, 74 per cent of seed treatment and 66 per cent of soil treatment and 42 per cent about spray of sulphur. The data shows that the respondents were mostly aware about the curative measures and less aware about preventive measures like soil treatment and seed treatment except the practice, seedling root shoot dip to which 90 per cent respondents were aware. The reason for this might be the compatibility of this practice with the prevailing practice of dipping the roots of the seedlings before transplanting in the solution of common salt and *Hing*. It is also worthwhile to note that 58 per cent respondents were not aware of the spraying of sulphur solution to control downward leaf-curl which is caused by the mites. Most of the farmers were misunderstanding that downward leaf-curl was not curable and the cause might be any insect or disease germ but not the mites. Unawareness of the casual organism and wrong assumption might be the cause of unawareness about the sulphur spraying. Another factor might be the non-availability of the sulphur with the plant protection dealers might have also affected awareness in this respect. Likewise, non-availability of insecticides and less communicability of the results of preventive measures which are not quickly visible as in the curative measures within short time might have affected an awareness of the preventive measures like seed treatment and soil treatment. Therefore, in order to accelerate the usage of preventive measures and the usage of sulphur, more intensive extension work to educate the farmers appears to be needed, together with the use of mass media by the state machinery as well as private agencies concerned.

4.2.2 Adoption

The Table 6 presents the adoption of recommended plant protection practices by the respondents. About 82 and 52 per cent farmers adopted spraying of insecticide plus

Table 6 : Adoption of recommended plant protection measures by the respondents

Sr. No.	*Recommended plant protection measures*	*Adoption of plant protection measures*		
		Number	*Percentage*	*Rank*
1.	Spray of insecticide plus pesticide	164	82.00	I
2.	Seedling root shoot dip	148	74.00	II
3.	Spray of insecticide plus pesticide after flowering.	104	52.00	III
4.	Seed treatment.	100	50.00	IV
5.	Spray of sulphur	84	42.00	V
6.	Soil treatment	56	28.00	VI

pesticide before and after flowering respectively followed by 74 per cent seedling root-shoot dip, 50 per cent seed treatment, 42 per cent spraying of sulphur and the least soil treatment, i.e. 28 per cent in descending order.

4.2.3 Awareness-adoption gap

The Table 7 indicates that awareness adoption gap is highest in case of soil treatment, i.e., 56 per cent followed by spraying of insecticide and pesticide after flowering (46%), seed treatment (32%) spray of insecticide plus pesticide before flowering and seedling root-shoot dip (18% each, respectively). It is worthwhile to note that awareness-adoption gap with regard to spraying of sulphur, the valuable practice to control downward leaf-curl which causes the intensive loss to the chilli growers; the farmers who were aware about this practice and knowing the cause and effect, all of them had adopted this practice and awareness-adoption gap was zero.

Table 7 : Farmers' awareness adoption gap about the recommended plant protection measures

Sr. No.	*Recommended plant protection measures*	*Awareness No. of farmers*	*Adoption No. of farmers*	*Awareness Adoption gap No. of farmers*	*Percentage*	*Rank*
1.	Soil Treatment	132	56	76	56.00	I
2.	Spray of insecticide plus pesticide before flowering	192	104	88	46.00	II
3.	Seed treatment	148	100	48	32.00	III
4.	Spray of insecticide plus pesticide after flowering	200	164	36	18.00	IV
5.	Seedling root-shoot dip	180	148	32	18.00	IV
6.	Spray of sulphur	84	84	0	0.00	V

The spraying of insecticide and pesticide before flowering and seedling root-shoot dip in which awareness-adoption gap was narrow, i.e. 18 per cent in each case. It is the fact that most of the dryland farmers do not take any plant protection measures for the other crops except cash crop. Chilli being grown as a cash crop, at least respondents in the present investigation took spraying of insecticide plus pesticide before flowering to check the most important diseases, i.e., churda-murda (leaf curl) and dieback and the awareness-adoption gap was narrow, the same gap, i.e. 18 per cent has been observed in case with the seedling root-shoot dip which is also narrow because of the compatibility of this practice with the prevailing local pratice among the farmers to dip seedling root and shoot in the solution of Hing and common salt.

But as regard with the awareness-adoption gap of the practice, spraying of insecticide and pesticide after flowering which more i.e. 46 per cent might be due to lack of capital and money etc., during the peak period of farm operations, i.e., from the month September onwards in which farmers generally diversify their resources inclusive money on all farm operations.

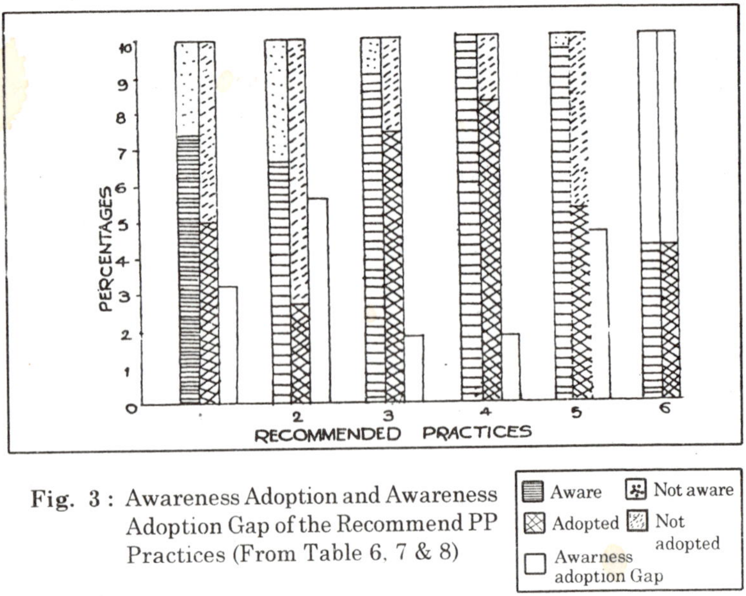

Fig. 3 : Awareness Adoption and Awareness Adoption Gap of the Recommend PP Practices (From Table 6, 7 & 8)

In case of soil treatment and seed treatment, awareness adoption gap was wide i.e., 56 per cent and 32 per cent respectively. It might be due to lack of visible effects of the practices which are of preventive nature to check the occurrence of insect pests and disease attack in future. In these case, only awareness might not have served the purposes, to convert the message into practice, i.e., adoption for which deep understanding about the importance of preventive measures should be known to the respondents as a result of extension efforts.

4.2.4 Frequency of sprayings of insecticides plus pesticide

The Table 8 indicates that the respondents did not follow the recommended numbers of spraying of insecticides plus pesticide before and after flowering as such. They modified the recommendations as per their need, socio-economic status and different perceived requirement at individual level. Out of 164 respondents who adopted the practice of insecticide plus pesticide before flowering; 39 per cent of them preferred to take only one spraying, 32 per cent three sprayings and 29 per cent two sprayings.

Table 8 : Frequency of insecticide plus pesticide spraying taken by farmers before and after flowering of chilli crop

Sr. No.	*No. of sprayings*	*Before flowering N=164*		*After flowering N=104*	
		Number	*Percentage*	*Number*	*Percentage*
1.	3 sprayings	52	32.00	24	23.00
2.	2 sprayings	48	29.00	12	12.00
3.	1 sprayings	64	39.00	68	65.00

With regard to the spraying of insecticide plus pesticide after flowering; out of 104 farmers, majority of them i.e., 65 per cent preferred to take only one spraying as against 23 per cent three sprayings and 12 per cent two sprayings.

From the above facts, it can be clearly inferred that farmers preferred to spray once each before and after flowering. It might be due to increase in the cost of production of crop which may not recover after selling of produce, being the low prices to the chillies in the market. Fear of risk and uncertainity also played an important role to adopt the number of sprayings before and after flowering. Non-availability of water for spraying after the month of September resulted to take forcibly only the dusting operations as stated by 20 per cent respondents.

4.2.5 Level of Adoption

The Table 9 clearly shows that with respect to all the plant protection measures near about half and above half of the respondents used recommended doses of the

Table 9 : Level of adoption of the recommended doses of plant protection measures taken by farmers

Sr. No	*Plant protection Measures*	*No.of farmers*	Adoption level					
			Recommended		*Above Recommended*		*Below Recommended*	
			No.	*dose percentage*	No.	*dose percentage*	No.	*dose Percentage*
1.	Seed treatment	100	60	60.00	32	32.00	8	8.00
2.	Soil treatment	56	48	86.00	4	7.00	4	7.00
3.	Seedling root shoot dip	148	72	48.00	48	33.00	28	19.00
4.	Spray of insecticide plus pesticide before flowering	164	84	51.00	8	5.00	72	44.00
5.	Spray of insecticide plus pesticide after flowering	104	44	42.00	8	8.00	52	50.00
6.	Spray of sulphur	84	40	48.00	--		44	52.00

chemicals. It also shows that near about half of the respondents used below recommended doses as in case with the curative measures such as spraying of sulphur, insecticide plus pesticide spraying before and after flowering as against very less majority of the respondents used below recommended doses with regard to the preventive measures' practices such as seedling root shoot dip (19%), seed treatment (8%) and soil treatment (7%).

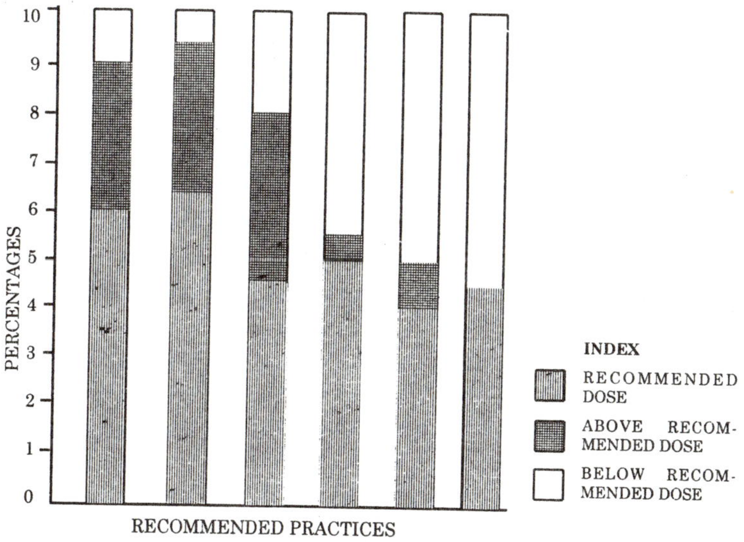

Fig. 4: Level of adoption of the Recommended doses of PP measures taken by the respondents (from Table 10)

It is also observed that with regard to the preventive measures like seed treatment and seedling root-shoot dip, the farmers used them above recommended doses except in case with the soil treatment in which cost involvement is more. Whereas, in case with the curative masseurs they hardly used them above recommended doses. This tendency might be due to the less quantity of chemical required for the use of preventive measure practices which involve low cost, whereas in case with the curative measures, the quantity of the chemical requires more which involve more cost.

From the foregoing discussion, it can be clearly stated that the plant protection measures which involved more cost, farmers used them by following below recommended doses and the plant protection measures which involved less cost to practice, farmers used them by following the recommended doses.

4.2.6 Correct Stage/time of the Plant Protection Measures taken by the Dryland Farmers

It is evident from the Table 10 that majority of the respondents followed plant protection measures at the right time with regard to the preventive measures such as seed treatment, soil treatment and seedling root-shoot dip with corresponding 92, 93 and 97 per cent respondent respectively. But, in case with the curative measures the percentage of the respondents in each case is less as compared to preventive measures' practices.

Table 10 also shows that farmers followed plant protection measures even before the right stage or time though not required. It happened particularly in case with the curative measures rather than preventive measures. Further, it reveals that 37, 27 and 62 per cent respondents respectively, followed plant protection measures before the right time or stage as in case with the curative measures such as spraying of insecticide plus pesticide before as well as after flowering and spraying of sulphur respectively as against very meagre percentage observed in case with the preventive measures. Incidently 2 per cent and 5 per cent respondents respectively followed the practice of spraying of insecticide plus pesticide before flowering and spraying of sulphur after the right stage or time.

4.2.7 Reinvention in the use of Plant Protection Measures by the Dryland Farmers

The Table 10 shows that the farmers modify or reinvent the plant protection practices to suit their requirement. The reinvention was more in plant protection practices which involve high cost. In this case the farmers used inputs below the recommended doses. Inversely, in the plant protection practices involving low cost; the farmers have the tendency to adopt the recommended doses. Thus,' most

Table 10 : Stage/time of undertaking the plant protection measures by the farmers

Sr. No.	*Recommended plant protection measures*	*No. of farmers*	*Stage/time of plant protection measures*					
			No.	*At right stage*		*Before right stage*		*After right stage*
				percentage	*No.*	*percentage*	*No.*	*percentage*
1.	Seed treatment	100	92	92.00	8	8.00	—	—
2.	Soil treatment	56	52	93.00	4	7.00	—	—
3.	Seedling root - shoot dip	148	144	97.00	4	3.00	—	—
4.	Spray of insecticide plus pesticide before flowering	164	100	61.00	60	37.00	4	2.00
5.	Spray of insecticide plus pesticide after flowering	104	76	73.00	28	27.00	—	—
6.	Spray of sulphur	84	28	33.00	52	62.00	4	5.00

of the farmers have adopted seed treatment and seedling root-shoot dip, practices. It is also worthy to note that the progressive chilli growing farmers followed the recommended doses of plant protection chemicals as such.

Reinvention occurs not only in case with the recommended doses of the chemicals but with the frequency of spraying applications as it is evident from Table 8. It showed that majority of the respondents, that is, 39 per cent and 65 per cent respectively preferred to take only one spraying of insecticide plus pesticide before flowering and one sprayings after flowering as against total of six recommended sprayings. Non-availability of water for spraying after the month of September resulted to take forcibly only the BHC dusting operation as stated by 20 per cent farmers. It is, also observed that more than 80 per cent respondents followed the interval of more than 20 days between two sprayings as against recommended interval of 10 days.

It is evident from Table 10 that the farmers used chemicals haphazardly even before the right stage or after the right stage of pest attack. They did not think about economic threshold level of the injury of insects pests and disease attack. Lack of appropriate knowledge of plant protection practices, unawareness about the resultant consequences of chemicals' application in haphazard manners, negligence and non-seriousness of the dryland farmers in handling sophisticated technology like plant protection seemed to be the reasons behind reinvention in the use of plant protection measures.

The foregoing discussion showed that the dryland farmers modify the recommendations in case with the recommended doses, frequency of sprayings to be followed, the interval between two spraying etc., as per their need socio-economic status and different perceived requirements at the individual level. It was also observed that other fellow farmers followed or imitated unknowingly or knowingly as the progressive farmers followed on their fields.

4.3. FACTORS AFFECTING PLANT PROTECTION STATUS OF DRYLAND FARMERS

4.3.1 Inter-correlation matrix of the variables selected for study

To get a comprehensive idea of the intercorrelation between and among the independent variables and the dependent variable, the intercorrelation matrix analysis was done and the results are presented in Table 11.

A close examination of inter-correlation matrix presented in Table 11 shows that out of 105 combinations ($15 \times 14/2 = 105$) of correlation coefficients, 92 were positive and 13 were negative.

Table 11. Inter-correlation matrix of the variables selected for study. (N=200)

Variable No.	*Name of the variables*	*Inter-correlation matrix*														
		X_{01}	X_{02}	X_{03}	X_{04}	X_{05}	X_{06}	X_{07}	X_{08}	X_{09}	X_{10}	X_{11}	X_{12}	X_{13}	X_{14}	X_{15}
X_{01}	Age	1.0000 **														
X_{02}	Education	-0.4594 **	1.0000 NS													
X_{03}	Family size	0.2129 **	0.0954 **	1.0000 **												
X_{04}	Land holding	-0.2167 NS	0.4312 **	0.3767 **	1.0000 **											
X_{05}	Social participation	-0.1340 **	0.6949 **	0.2259 **	0.5268 **	1.0000 **										
X_{06}	Socio-economic status	**	-0.2292 **	0.7144 **	0.4004 **	0.8527 **	0.7382 **	1.0000								
X_{07}	Availability of PP equipments	-0.2318 NS	0.4047 **	0.2176 **	0.5070 **	0.4533 **	0.5766 **	1.0000 **								
X_{08}	Agricultural infrastructure	-0.0888 NS	0.4329 **	0.3064 **	0.6553 **	0.5010 **	0.6170 **	0.4796 **	1.0000 **							
X_{09}	Cropping intensity	-0.0999 NS	0.5330 **	0.2527 *	0.5220 **	0.5220 **	0.5366 **	0.3315 **	0.5612 **	1.0000 **						
X_{10}	Economic motivation	-0.1201 NS	0.6374 **	0.2012 *	0.4554 **	0.6266 **	0.6061 **	0.4843 **	0.4726 **	0.4684 **	1.0000 **					
X_{11}	Scientific orientation	-0.1573 NS	0.6081 **	0.2059 **	0.4346 **	0.5577 **	0.5294 **	0.4664 **	0.5251 **	0.4436 **	0.8495 **	1.0000 **				
X_{12}	Risk preference	-0.326 *	0.5756 **	0.3563 **	0.4162 **	0.5222 **	0.5346 **	0.3983 **	0.4871 **	0.4503 **	0.8512 **	0.8938 **	1.0000 **			
X_{13}	Utilization of sources of information	-0.1662 *	0.6405 **	0.3086 **	0.6739 **	0.5985 **	0.8115 **	0.5294 **	0.586 **	0.4703 **	0.6563 **	0.7140 **	0.6501 **	1.0000 **		
X_{14}	Adoption	-0.0039 *	0.6198 **	0.3212 **	0.5804 **	0.5199 **	0.6842 **	0.4655 **	0.5945 **	0.3160 **	0.5644 **	0.6555 **	0.6560 **	0.8196 **	1.0000 **	
X_{15}	Plant protection status	-0.1607	0.6981	0.2409	0.6287	0.5733	0.6993	0.4249	0.6073	0.4580	0.6773	0.7165	0.7335	0.8500	0.7831	1.0000

* Significant at 5 per cent level ** Significant at 1 per cent level NS =Not Significant

Table 11 also shows that out of 92 positive combinations, 89 of them were positively significant at 1 per cent and 2 at 5 per cent level, whereas one combination of family size with education has shown non-significant result.

Further, the table reveals that out of 13 negative combinations, 4 were negatively significant at 1 per cent level and 3 at 5 per cent level; whereas 6 combinations have shown non-significant results.

Correlation coefficients of age with almost all variables were negative except with education which is positive.

Thus, it is clear from the inter-correlation matrix that all the variables undertaken for the study are significantly correlated with each other, between and among the independent variables and the dependent variable. Hence, a set of independent variables selected for studying the plant protection status of the dryland farmers is quiet concrete to see the further relationship, contribution and their direct and indirect effects.

In a bid to explore the factors that affect the plant protection status of the dryland farmers, relational analysis were conducted and the results are shown in the forthcoming tables.

4.3.2 Relationship between independent variables and plant protection status level of the dryland farmers

The computed coefficient of correlations (r) in Table 12 indicates that except age all the independent variables are related positively and significantly with the plant protection status level of the dryland farmers and the results are significant at 0.01 level of probability. This clearly reveals that, increase in scores of these variables would increase the plant protection status level of the dryland farmers.

Age has shown the negative and significant relationship ($r= -0.160$) with plant protection status. It means as the age of the dryland farmers increases, plant protection status level decreases.

Table 12 : Relationship between independent variables and plant protection status of the respondents

Variable No.	*Name of the variables*	*'r' values*
X_1	Age	0.1607 *
X_2	Education	0.6981**
X_3	Family size	0.2409**
X_4	Land holding	0.6287**
X_5	Social participation	0.5733**
X_6	Socio-economic status	0.6993**
X_7	Availability of plant protection equipment	0.4249**
X_8	Agricultural infrastructure	0.6073**
X_9	Cropping intensity	0.4580**
X_{10}	Economic motivation	0.6773**
X_{11}	Scientific orientation	0.7165**
X_{12}	Risk preference	0.7335**
X_{13}	Utilization of sources of information	0.8420**
X_{14}	Adoption	0.7831**

* Significant at 0.05 level of probability
** Significant at 0.01 level of probability

The Table 12 also shows that the independent variables such as utilisation of sources of information, adoption risk preference, scientific orientation have more 'r' values i.e., 0.8420, 0.7831, 0.7335, 0.7165 respectively, which state the positive and significant relationship with plant protection status of the respondents in descending order.

Next independent variables having significant and positive relationship with plant protection status are socio-economic status, education, economic motivation, land holding, agricultural infrastructure, social participation, cropping intensity and availability of plant protection equipments with corresponding 'r' values 0.6993, 0.6981, 0.6773, 0.6287, 0.6073, 0.5733, 0.4580 and 4249 respectively in descending order. As compared to all other independent variables family size and age have shown very less 'r' values, that is 0.2409 and -0.1607 respectively.

Education played important role in enhancing technical know-how. It helped in widening the mental horizon of an individual farmer. Technical literature produced by the various agencies involved in dissemination of technical know-how might have been read by the educated farmers and some of it, they brought into practice. Unlike educated person, person with no formal education, encounters a number of problems and difficulties in understanding techniques involved in plant protection. It may be due to this reason that education (r=0.6981) was positively and significantly correlated with plant protection status.

Family size (r=0.2409) showed the positive and significant relationship with plant protection status. This shows that in the peak period of the season, more number of family members in big size family facilitated to held the responsibilities in relation with the plant protection work in absence of hired labour, as a result plant protection status level would be high due to increased interaction which resulted to get higher knowledge, attitude, skill and adoption, ultimately led towards high plant protection status.

Land holding is one of the criteria, which not only decides the standard of living and prestige of an individual in the society, but as an agriculturist having more lane, acts as an incentive towards increase the production by accepting scientific agricultural practices including plant protection which ultimately helps to achieve high plant protection status.

High Social participation enabled the farmers to come into contact with various sources of information which in turn increased their level of attitude, knowledge, skill and adoption of plant protection practices, which ultimately led towards high level of plant protection status.

In the present study, apart from other items, socio-economic status scale includes some of the important items such as education, socio-political participation, land holding, farm power/ possession etc. as an individual's factors which enabled the respondents to come into contact with reliable sources of information which helped them to increase their knowledge, change their conviction pattern to bring about increased adoption of plant protection practices and ultimately the plant protection status. Sen and Doijad[74] also supported the above finding.

Farmers having high agricultural infrastructure facilities will naturally try to utilise the available resources at the maximum extent by exploring all the facilities by way of increasing the high cropping intensity for getting the maximum yield. This tendency enabled the farmers in achieving more knowledge and skill about the technology which helped to increase the profit by way of adopting the technology like plant protection on their farms which ultimately led towards high plant protection status of the respondents.

The data from Table 12 showed that psychological variables like risk preference (r= 0.7335), scientific orientation (r=0.7165) and economic motivation (r = 0.6773) were found to be positively and significantly related with the plant protection status of the respondents. It is natural that large investment is needed for plant protection and this investment in made by the farmers who are ready to take risk. This might be the reason for significant relationship between the risk preference and plant protection status of the respondents. It is also logical that the farmers who have faith in scientific knowledge (like plant protection) have better plant protection status. The economically motivated farmers are aware of the input-output relationship and consider the investment in plant protection as beneficial. In view of this there might have been significant relationship of this variable with plant protection status.

4.3.3 Multiple regression analysis

It is evident from the Table 13 that all the independent variables have shown significant relationship with the plant protection status of the farmers. Therefore, in the first step all the 14 independent variables have been considered for multiple regression analysis with a view to determine the contribution of all the independent variables in predicting the plant protection status of the dryland farmers. B values (partial regression coefficients) of all the independent variables with plant protection status are presented in Table 13.

Table 13 : Regression coefficients of plant protection status with the independent variables

Variable No.	*Name of the variables*	*B value (RC)*	*'t' value for (B)*
X_1	Age	0.0778	0.7049
X_2	Education	6.5347	5.4728*
X_3	Family size	-25583	1.5332
X_4	Land holding	7.7499	4.7047*
X_5	Social participation	1.3056	1.2940
X_6	Socio-economic status	-1.9672	4.6825*
X_7	Availability of plant protection equipment	-11.5222	4.5883*
X_8	Agri. infrastructure	0.8354	4.2924*
X_9	Cropping intensity	-0.1566	2.9149*
X_{10}	Economic motivation	0.1703	0.5089
X_{11}	Scientific orientation	-0.2379	3.7611**
X_{12}	Risk preference	1.6938	5.3512*
X_{13}	Utilization of sources of information	1.9820	9.9221*
X_{14}	Adoption	0.0181	0.1674

* Significant at 0.01 level of probability.
R^2=85.59 F = 78.50

Table 13 reveals that out of 14 independent variables, the regression coefficients of education (b=6.5547) land holding (b=7.7499), socio-economic status (b=-1.9672) availability of plant protection equipment (b-11.522), agricultural infrastructure (b=0.3354), cropping intensity (b=-0.1566) scientific orientation (b=-0.2375), risk preference (b=1.6936) and utilization of sources of information (b=-1.9820) are significant. These nine variables may be termed as good predictors in influencing the plant protection status of the dryland farmers.

The value (R^2) suggests that all the independent variables jointly contributed 85.59 per cent towards variation in plant protection status. The F value (F=78.50) was also found to be significant at 0.01 level of probability indicating the significant effectiveness of these variables in influencing the plant protection status of the dryland farmers, when they function jointly.

In the second step, the step down regression analysis was computed to determine the contribution of the significant variables eliminating the other non-significant variables.

Table 14 : Regression coefficients of plant protection status with the significant independent variables

Variable No.	*Name of the variables*	*B value (RC)*	*'t' value for (B)*
X_2	Education	6.9037	7.3426
X_4	Land holding	6.6496	4.2867
X_6	Socio-economic status	-1.7422	5.0896
X_7	Availability of plant protection equipments	-11.3849	4.7659
X_8	Agri. infrastructure	0.8327	4.6347
X_9	Cropping intensity	-0.1419	2.9364
X_{11}	Scientific orientation	-0.9882	3.4070
X_{12}	Risk preference	1.5739	6.2697
X_{13}	Utilization of sources of information	1.9147	12.7121

R^2=85.07 F=78.50

Table 14 reveals that the regression coefficient of education (b=0.9037), land holding (b=6.6396), socio-economic status (b=-1.7422), availability of plant protection equipment (b= 11.3849), agricultural infrastructure (b= 0.8327),

Table 15 : Path coefficents (analysis) showing the direct and indirect effect if ubdeoebdebt variables on the plant protection status of the respondents

Vari-bles No.										
X_{01}	Age	-0.1607	-0.0294	-0.1313	0.1441	X_{6}	-0.1337	X_{13}	0.1319	X_{02}
X_{02}	Education	0.6981	0.2871	0.4110	0.5153	X_{13}	-0.3532	X_{06}	0.2666	X_{12}
X_{03}	Family size	0.2409	-0.0486	0.2895	0.2483	X_{13}	-0.1979	X_{06}	0.1650	X_{12}
X_{04}	Land holding	0.6287	0.2975	0.3312	0.5422	X_{13}	-0.4216	X_{06}	0.1927	X_{12}
X_{05}	A social participation	0.5733	0.0897	0.4836	0.4815	X_{13}	-0.3649	X_{06}	0.2418	X^{12}
X_{06}	Socio-economic status	0.6993	-0.4944	1.1937	0.6529	X_{13}	0.2537	X_{04}	0.2473	X_{12}
X_{07}	Availability of PP equipment	0.4249	-0.1930	0.6179	0.5064	X_{13}	-0.2851	X_{06}	0.1845	X_{12}
X_{08}	Agril. Infrastructure	0.6073	0.2014	0.4059	0.4655	X_{13}	-0.3050	X_{06}	0.2256	X_{12}
X_{09}	Cropping intensity	0.4580	-0.1203	0.5783	0.3784	X_{13}	-0.2653	X_{06}	0.2085	X_{12}
X_{10}	Economic motivation	0.6773	0.0452	0.6321	0.5280	X_{13}	0.3942	X_{12}	-0.2996	X_{06}
X_{11}	Scientific orientation	0.7165	-0.3440	1.0605	0.5744	X_{13}	0.4139	X_{12}	-0.2617	X_{06}
X_{12}	Risk preference	0.7335	0.4631	0.2704	0.5311	X_{13}	-0.3075	X_{12}	-0.2643	X_{06}
X_{13}	Utilisation of sources of information	0.8420	0.8045	0.0375	-0.4012	X_{6}	0.3057	X_{12}	-0.2456	X_{11}
X_{14}	Adoption	0.7831	-0.0214	0.8045	0.6594	X_{13}	-0.3383	X_{06}	0.3038	X_{12}

cropping intensity (b=-0.1419), scientific orientation (b=0.9882), risk preference (b=1.5739) and utilisation of sources of information (b=1.9147) are significant and jointly contributed 85.07 per cent variation in influencing the plant protection status. The F value (F=78.50 also found to be significant at 0.01 level of probability. This clearly confirm that these 9 variables are the important one in influencing the plant protection status i.e., upto 85.07 per cent.

4.3.4 Path Analysis

With a view to have a clear understanding about the channelising influence of the significantly correlated variables with the plant protection status, path analysis was attempted and results presented in Table 15 and diagramatically presented in Fig.5.

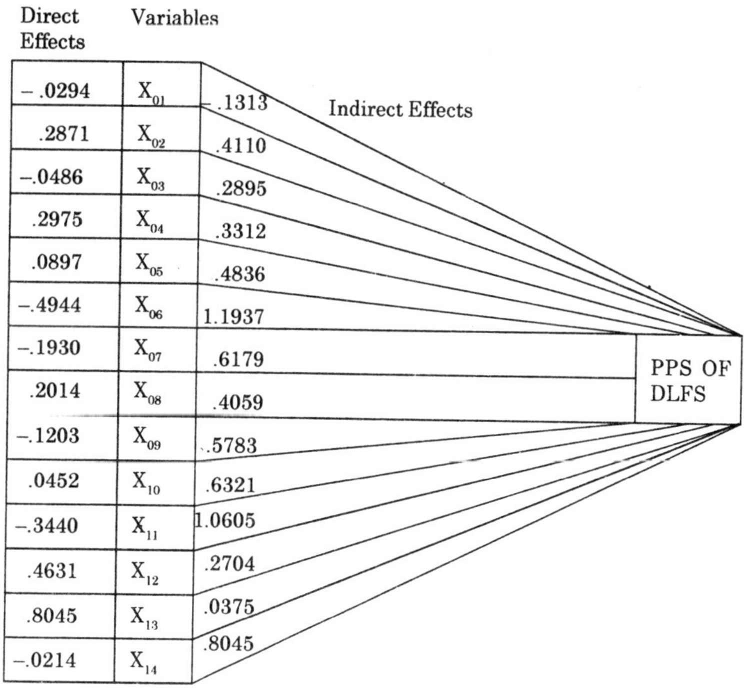

Fig. 5 : Path Diagram showing Direct and Indirect effects of Independent variables of PPS of DLDS (from Table 16)

A glance at Table 15 showed that all the independent variables undertaken for the study influenced the plant protection status directly as well indirectly in positive as well as in negative direction. The maximum direct effect on the plant protection status was exerted by the utilization of sources of information (0.8420.) Second in order of importance were socio-economic status (–0.4944) and risk preference (0.4631). After these variables, next in order of importance were scientific orientation (-0.3440), land holding (0.2975), education (0.2871), agricultural infrastructure (0.2014) and availability of plant protection equipment (–0.1930) and cropping intensity (–0.1203). Indirectly, socio-economic status found to be affecting the plant protection status to a very high extent (1.1937) and much it was channelised through utilization of sources of information followed by land holding and risk preference. Scientific orientation found to be second in order of importance which affect indirectly (1.0605) the plant protection status and much it was channelised through the utilization of sources information followed by risk preference and socio-economic status. Adoption found to be third in position showing the indirect effect (0.8045) and much it was channelised through utilisation of sources of information followed by socio-economic status and risk preference.

In the entire gamet of channelising the substantial indirect effect through other variables selected for study; it was observed that at twelve places each; sources of information, socio-economic status and risk preference had shown important role. This clearly shows that in order of importance, the variables utilization of sources of information; socio-economic status and risk preference are very important variables not only because of their high direct effect (0.8045, –0.4944, 0.4631 respectively on plant protection status) but also due to their role in channelising the substantial indirect effect through different variables in influencing plant protection status of dryland farmers. The variables, sources of information and socio-economic status had also found to be important in influencing indirectly on the plant protection status. So it could be clearly inferred that these three variables, namely, utilization of sources of information, socio-economic status and risk preference are the most crucial variables in influencing the plant protection status of dryland farmers.

4.4 PROBLEMS OF TRANSFER OF PLANT PROTECTION TECHNOLOGY AT THE FARMERS' LEVEL IN ADOPTION OF PLANT PROTECTION MEASURES AS PERCEIVED BY THE FARMERS

The distribution of the respondents with regard to the problems faced by the dryland farmers in adoption of the plant protection measures are given in Table 16.

Table 16 : Problems of transfer of plant protection technology at the farmers' level in adoption of plant protection measures as perceived by the respondents (N=200)

	Problems	*No. of respondents*	*Percentage*	*Rank*
I.	**Situational**			
I.	Non -availability of irrigation facilities which paralyses the whole system of farm operations including plant protection	120	60.00	I
2.	Risk and uncertainities due to occurrence of pests and disease attack resulted to low profit	90	45.00	II
3.	Fear of more or low rainfall during the season which ultimately loose the confidence in loan repayment due to uncertainity in yield.	60	30.00	III
4.	Rain at the time of chilli harvesting resulted to uncertainity of market prices and fear of less benefit as compared to cost.	42	21.00	IV

Table 16 *(contd.)*

5.	Non-availability of irrigation facilities which paralyse the whole system of farm operations including plant protection.	40	20.00	V
6.	Water not available after the month of September for spraying and only dusting measures are forcibly taken.	20	10.00	VI
7.	Non-availability of skilled labour in plant protection operations at the time of peak period of the season affecting the usage of plant protection measures	19	9.05	VI
II.	**Economical**			
1.	High wages of the trained labour for plant	72	36.00	I
2.	Diversification of power and income during the kharif which is peak period which ultimately affected plant protection measures	70	35.00	II
3.	High cost of sprayers and dusters	60	30.00	III
4.	High cost of chemicals, pesticides and insecticides in kharif season	41	20.05	IV
5.	Benefit is low as compared to cost of cultivation including plant protection due to low market prices for green as well as dry chillies	35	17.05	V
6.	Lack of finance affect the plant protection measures.	28	14.00	VI
7.	A traditional way of thinking lead to routine way of life rather than use of plant protection.	10	5.00	VII
III.	**Technological**			
1.	Non-availability of churda-murda and other disease resistant varieties	120	60.00	I
2.	No fool proof research to control pest and disease	82	41.00	II
3.	Available technology is sophisticated and complicated as compared to knowledge of dryland farmers	52	26.00	III

Table 16 *(contd.)*

4.	Non-availability of suitable location specific plant protection technology for dryland farmers.	40	20.00	IV
5.	Non-availability of low cost sprayers and dusters suitable to dryland farmers	34	20.00	IV
IV.	**Technical problems at personal level**			
1.	Lack of technical knowledge of control measures of pests and diseases	89	44.05	I
2.	Lack of knowledge about the type of chemical usage and their concentration	78	39.00	II
3.	Lack of technical skill in preparation of concentration of chemical solution	69	34.05	III
4.	Lack of skill in repairing of plant protection equipment	59	29.05	IV
5.	Lack of knowledge of Govt. schemes for loans, subsidy etc. for taking plant protection measures	50	25.00	V
6.	Lack of knowledge about appropriate time of application of chemical at the particular stage of the crop growth or pest and disease stage.	49	24.05	VI
7.	Lack of technical skill in handling of plant protection equipment for chemical application	40	20.00	VII
8.	Lack of mobility in other chilli growing areas to see how other farmers manage the pests	30	15.00	VIII
V.	**Supply**			
1.	Lack of facilities of obtaining sufficient plant protection equipment on loan basis	89	44.05	I
2.	Non-availability of sprayers and dusters in the peak period of spraying on hire basis	68	34.00	II
3.	Lack of availability of the required chemicals within reach of the farmers	62	31.00	III

Table 16 *(contd.)*

4.	Lack of efficient system to deliver pesticides, insecticides before the season.	48	24.00	IV
VI. Technical guidance and services				
1.	Lack of repairing facilities for sprayers and dusters nearby.	60	30.00	I
2.	Lack of mobility of extension worker in the field	40	20.00	II
3.	Lack of reliable sources of information in the sophisticated plant protection technology at the village level.	30	15.00	III
4.	Village level worker does not deliver the messages to the farmers in time	22	11.00	IV
5.	Non-availability of trained mechanic in the village to repair plant protection equipment in the field situations	20	10.00	V
6.	Weak support from information disseminating system	20	10.00	V
7.	Lack of required technical competence of village level worker at the field level	20	10.00	V
8.	Low level of communicational ability of extension worker	14	7.00	VI
9.	Printed literature from the Department of Agriculture do not reach farmers	14	7.00	VI

4.5 EMPIRICAL MODEL

On the basis of empirical evidences and statistical tests and in the light of the above discussion of the results of present study, the most important variables exerting high significance and contribution in predicting influencing plant protection status have been scrutinized along with the problems faced by the dryland farmers in transfer of plant protection technology at the farmers' level. The final form of empirical model has been presented in Fig. 6.

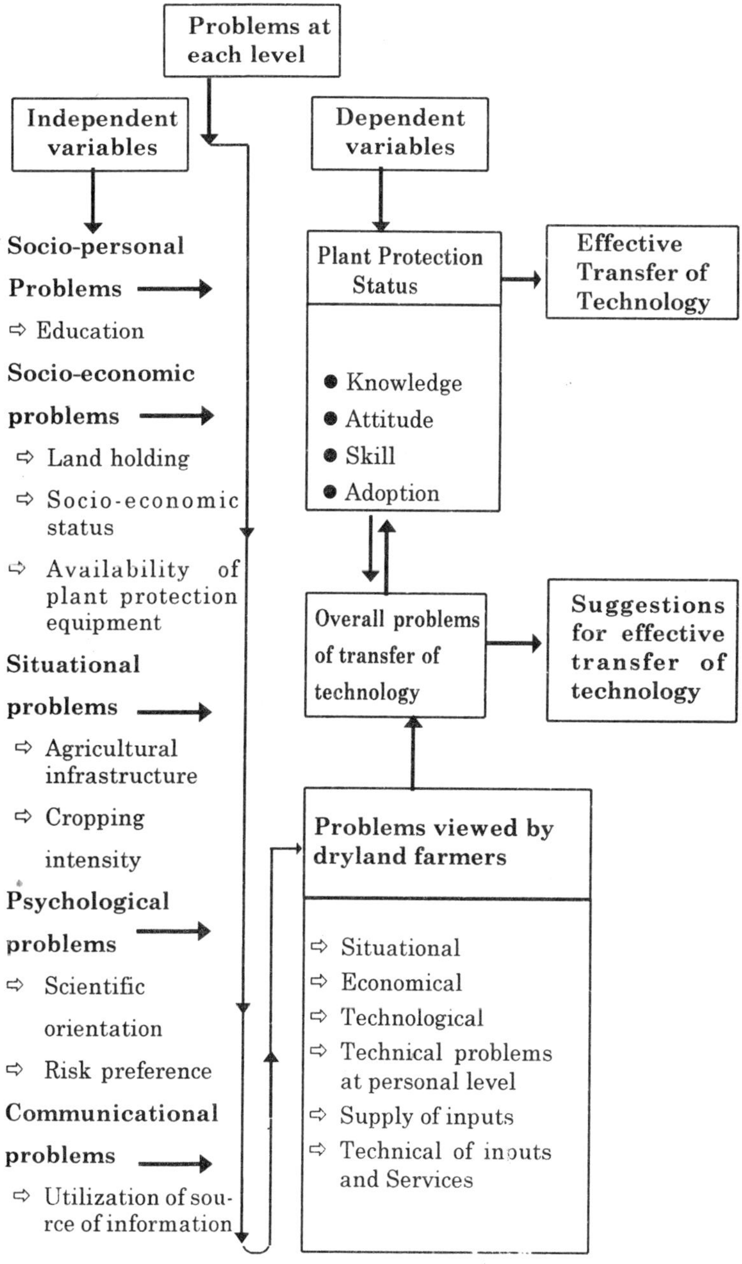

Fig. 6 : Empirical Model on problems of transfer of Plant protection technology among the dryland farmers

A comprehensive set of 14 independent, variables was undertaken for study with a view to establish their relationship with plant protection status of the dryland farmers as shown in the conceptual model (Fig.1.). Out of these 14 variables only 9 variables i.e. education, land holding, socio-economic status, availability of plant protection equipment, agricultural infrastructure, cropping intensity, scientific orientation, risk preference and utilisation of sources of information have been found highly significant and also have significant contribution for variation in the plant protection status of the dryland farmers. All these 9 characteristics of an individual dryland farmer which are categorised under socio-personal, socio-economic, situational, psychological and communicational act as problem-complex which affect the plant protection status and ultimately the transfer of plant protection technology on the farmers' field. Practical problems as viewed by an individual dryland farmer which have been categorised under situational, economical, technological, technical problems at personal level, problems at supply level, and problems of technical guidance and services as a whole along with above mentioned nine characteristics depicts the overall problem-complex for effective transfer of plant protection technology at the farmers' level in the dry farming areas. The different suggestions derived from the present investigation (see chapter-V 'Implications') may prove useful in formulating a suitable strategy for effective transfer of plant protection technology among the dryland farmers.

5

IMPLICATIONS

The following prominent implications have been emerged from the findings of the study which are of practical use to the developmental agencies who are engaged in promoting the plant protection work particularly in the dryland areas.

5.1.1. A comprehensive set of significant variables which contribute to achieve more plant protection status of dryland farmers may be considered in the transfer of plant protection technology

The positive and high significant relationship of independent variables such as education, land holding, socio-economic status, availability of plant protection equipment, agricultural infrastructure, cropping intensity, scientific orientation, risk preference and utilisation of sources of information with plant protection status and their significant and jointly contribution of 85.07 per cent towards variation in plant protection status of dryland farmers suggest that developmental agencies / extension staff can utilise these traits whenever involved in the plant protection activities.

The overall analysis shows the dominant influence of the variable, utilisation of sources of information on the plant protection status of the dryland farmers. It clearly indicates that all the sources of information categorised under mass media, personal cosmopolite and personal localite sources should be geared up for scientific orientation to accelerate the plant protection work as a result of extension efforts.

5.1.2. Credible sources from the point of view of dryland farmers may be utilised for popularising plant protection technology

Progressive farmers of the village were found to be the most credible source of information for the dryland farmers. The transfer of technology would be effective if the extension personnel communicate the plant protection technology through the progressive farmers of the village, since dryland farmers trust this source more. It is also suggested that progressive farmers should be encouraged to participate actively in the extension activities and acquaint with more ideas about plant protection. The Village Level Extension Workers and Agriculture. Extension Officers were found to occupy the second and third position as the credible sources in the most used sources of information by the dryland farmers in getting the information about plant protection. Village Level Workers and Agriculture. Extension Officers should be well trained in the latest plant protection concepts to coup up with the technological demands of the dryland farmers in connection with the plant protection. Based on pattern of utilisation of sources of information suitable extension strategy can be formulated to accelerate the plant protection work.

5.1.3. Concentration of efforts would be required to create awareness about plant protection technology among dryland farmers

Scientists and extension workers should do all the concentration of efforts for effective transfer of plant protection technology. Scientists/Extension workers should write comprehensively on the latest plant protection technology and publish it through University or State Department of Agriculture. The technology thus written by the scientists should be released through newspapers, radio and TV broadcast, publications in the form of leaflets, folders and bulletins extensively for the use of farmers.

5.1.4. Suitable strategy may be developed for risk coverage of the dryland farmers

Risk preference also emerged as the dominant variable which affects the plant protection status of the dryland farmers. In order to cover the risk, the following implications may be considered.

Insect pest and disease resistant varieties may be developed by the scientists against the serious insect pest and disease such as in Chillies, the resistant variety against Churda-murda may be the important necessity of the dryland farmers. A strategy may be developed to control pest and disease as full proof measure by the Scientists. Crop insurance scheme may be followed in the dryland areas. Reasonable market prices should be guaranteed from the Government side. Community nurseries should be raised and for this developmental agencies should take leading part for such activities. Crops may be sprayed or dusted whenever required by the Government agencies or contract should be given to the private agencies and the reasonable rates should be collected from the farmers in lieu of this. Campaigns may be arranged by the extension agencies against the important insect pest or disease such as Churda-murda, dieback etc. Necessary possibilities may be explored for life saving irrigation and necessary arrangements of water for spraying in the slack period of water availability may be made.

5.1.5. Facilities to supply plant protection equipment on loan basis and their repair and maintenance may be arranged by the extension agencies.

Plant protection equipments may be supplied on loan basis with subsidy on instalments to the farmers by an extension agency through cooperative societies or Gram

Panchayats on rent basis. The type of plant protection equipment available in the particular locality may be surveyed and trained mechanic may be appointed in such areas to repair those equipment. Such as in the present area of study foot sprayers, knapsack sprayers, hand rotary dusters and power sprayers are available. Orientation training for one or two days may be arranged for the extension workers, progressive farmers, other farmers and mechanics in that particular area to repair and maintain such equipment.

5.1.6. Suitable measures may be taken to support and improve the socio-economic status of dryland farmers

For all the above implications, it is suggested that the socio-economic condition of the dryland farmers may be improved. The sophisticated technology like plant protection requires favourable socio-economic condition of the farmers for bringing about their increased plant protection status level. Efforts may be made to improve the socio-economic status of farmers.

5.1.7. Special awareness about preventive as well as the most important plant protection measures may be created among the dryland farmers, highlighting the economic importance

Awareness may be created among the dryland farmers regarding importance of preventive measures in the entire cycle of plant protection. For most important practices, such as the present study, spraying of wettable sulphur to control downward leaf-curl, the root cause of the disease and the effect of the chemical etc., should be deeply understood by the farmers; for this extension training in plant protection is required to the dryland farmers. Some case studies and success stories may be published frequently and released time to time in the form of literature, highlighting the importance of preventive measures in plant protection.

Interviews of farmers may also be released through radio and TV programmes.

5.1.8. Provision of necessary training on Integrated Pest Management may be made to dryland farmers

Indiscriminate use of pesticides by the farmers results in problems like residue, resistance in insects, pollution in air, water and soil and deplection of natural enemies of injurious species. In view of the above hazards, due to indiscriminate use of pesticides, the present strategy of integrated pest management should be followed. This advocates the minimum use of pesticides that too should only be used at a certain level (i.e. economic threshold); it should not be used at all if injury (i.e. threshold of tolerance) does not justify the treatment. For this, suitable training programmes may be arranged for the dryland farmers, highlighting the importance of the above aspects in the use of plant protection technology.

APPENDICES

APPENDIX—I

INTERVIEW SCHEDULE FOR FARMERS PUNJABRAO KRISHI VIDYAPEETH, DEPTT. OF EXTENSION, POST GRADUATE INSTITUTE, AKOLA

Title of Research Project : Problems of transfer of technology of plant protection among dryland farmers of Vidarbha region of Maharashtra State.

Schedule No. : Date

Name of respondent : Village

1. Age
2. Education
3. Land holding (hectares)
4. Socio-economic status

Items	*Scores*
1. Occupation	
No	0
Occupation unshifted	1
Semi-shifted	2
Shifted	3

Farming business	4
Professional	5
2. ***Land holding (hectares)***	
Landless (No land)	0
Marginal (0.1-1.0)	1
Small (1.1-2.0)	2
Semi-medium (2.1 -4.0)	3
Medium (4.1 -10.0)	4
Large (10.1 and above)	5
3. ***Caste***	
Schedule	1
Most backward	2
Backward	3
Forward	4
Dominent	5
4. ***Education***	
No school education	0
Functional literal	1
Primary school	2
Middle school	3
High school	4
College	5
5. ***Socio-political participation***	
Without any position in Socio-Political Organisation	0
Official position in one or more organisation	1
Official position in social and Political Organisation	2

Financial contribution or raising funds for community work	3
Active office-bearer	4
Involvement in community	5

6. ***Possession***

None	0
One farm animal or material (Bullock, buffalo, cow, bicycle, furniture)	1
Two farm animals or materials (Bullock cart, radio)	2
Three farm animals or materials (improved farm implements, newspapers electricity)	3
Five to ten farm animals or materials (Gobar gas, spraying equipments, motor cycle)	4
More than 10 farm animals or materials (Tractor, automobiles, T.V)	5

7. ***House***

Shed thatched	1
Mud walls and thatched	2
Brick walls and tiled	3
Concrete house	4
Concrete and doubled storeyed	5

8. ***Household***

Small (1-3 members)	1
Medium (4-6 members)	2
Large (7-9 members)	3
Very large (9 and above)	4
Special features	5

5. Availability of PP Equipments

Sprayers -Yes/No

If yes, which sprayers__________

Dusters - Yes/No

If yes, which dusters____________

6. Agri. Infrastructural Facilities

(Scoring key: Very high-4, High-3, Average-2 unsatisfactory-1, very insignificant level-0)

	Very high	*High*	*Average*	*Unsat-isfactory*	*Very insign level*	*Score*
1. Irrigation						
2. Credit Institutions						
3. Electricity						
4. Input supply agencies						
5. Transportation						
6. Communication						
7. Media for Agril. Study						
8. Marketing facilities						
9. Land policy						
10. Storage facilities						

Total Score

7. Cropping Intensity (utilisation of land resources)

Net Area	*Area in acress*				*Cropping Intensity*
	Pre-kharif	*Kharif*	*Rabi*	*Gross*	

$$\text{Cropping intensity} = \frac{\text{Total annual cropped area (Gross)}}{\text{Size of holding in acress}} \times 100$$

Psychologial characteristics

(Scooring key - SA - 7, A - 5, UD - 4, DA - 3, SDA - 2
for - ve items scorring will be reverse)

8. Economic Motivation	*SA*	*A*	*UD*	*DA*	*S DA*	*Score*
1. A farmer should work toward large yield and economic profits.						
2. The most successful farmer is one who makes the most profit						
3. A farmer should try new farming idea which may earn him more money.						
4. It is important for the farmer to have a large harvest in order to be able to buy many things besides food						
5. It is difficult for the farmers children to make good start unless he can provide them economic assistance.						
6. It is not only monetary profit but the employment of work done, which give a farmer satisfaction for his hard work in farming						
Total Score						

9. Scientific Orientation *SA A UD DA S DA Score*

1. New methods of farming give better results to a farmer than the old methods
2. The way a farmer's forefathers farmed is still the best way to farm today
3. Even a farmer with lot of experience should use new method of farming
4. Though it takes time for a farmer to learn new methods in farming it is worth the efforts,
5. A good farmer experiments with new idea in farming
6. Traditional methods of farming have to be changed in order to raise the level of living of a farmer.

Total Score

10. Risk Preference

1. A farmer should grow large number of crops to avoid greater risk involved in growing one or two crops.
2. A farmer should rather take more of a chance in making a big profit than to be content with a smaller but less risky profits
3. A farmer who is willing to take greater risks than the average farmers usually does better financially.
4. It is good for a farmer to take risks when he knows his chance of success is fully high.
5. It is better for a farmer not to try new farming methods unless most of other farmers have used them with success.
6. Trying an entirely new method in farming by a farmer involves risk but it is worth it.

Total Score

11. Utilisation of Sources of Information

(Scooring key - key - most often - 4, Often - 3,
Sometimes - 2 rarely - 1, never - 0

Mass media	*Most often*	*Often*	*Some times*	*Rarely*	*Never*	*Score*
1. Radio broadcast						
2. Television broadcast						
3. Newspaper release						
4. Dept. publication (leaflet, folder, handouts etc).						
5. Farm literature (Shetkari magazine/ other magazines or books)						
Total Score						
Personal cosmopolite source						
6. Agril. Extension Officer						
7. VLW						
8. Bank personnels						
9. Pesticide dealer						
10. Progressive farmers of other villages						
Total Score						
Personal localite						
11. Neighbours						
12. Friends						
13. Relatives						
14. Progressive farmers of the village						
15. Village leaders						
Total Score						
Grand Total Score						

Source credibility

Put your first choice amongst all the above sources of information to get the information in the adoption of plant protection measures

12. Adoption of recommended plant protection measures and reinvention :

I. Seed treatment with Agrosan GN

1. Are you aware about seed treatment	Yes-1, No- 0
2. Did you treat the seed with Agrosan GN or any other chemical	Yes-1, No-0
3. How much quantity	Whole seed-3 more than half-2 Less than half-1
4. When treated	On the day of sowing-3 1 day before-2 2 day before-1
5. Rate per acre	Recommended dose-3 Above recommendation-2 Below recommendation-1

6. If not adopted or not adopted as such, what modified or alternative measures you have taken; specify reason.

Alternative /modification ______________________

reason ______________________

II. Soil Treatment with Phorate before Seed Sowing

1. Are you aware about soil treatment	Yes-1, No-0
2. Did you apply phorte in the seed bed	Yes-1, No-0
3. For how much area	Whole-3 More than half-2 Less than half-1

4. When did you apply	On the day of sowing-3 Before 1 day-2 After 1 day-1
5. Rate per acre	Recommended dose-3 Above recommendation-2 Below recommendation-1

6. If not adopted or not adopted as such what modified or alternative measures you have taken, specify reason.

Alternative/modification ______________________

reason ______________________

III. Seedling Root-shoot dip in DDT before Transplanting

1. Are you aware about above practice	Yes-1, No-0
2. Did you follow the above practice	Yes-1, No-0
3. Specify, the extent of seedlings treated	Whole-3 More than half-2 Less than half-1
4. When did you dip	Transplanting day-3 1 day before-2 2 days before-1
5. Rate per acre	Recommended dose-3 Above recommendation-2 Below recommendation-1

6. If not adopted or not adopted as such what modified or alternative measures you have taken, specify reason.

Alternative/modification ______________________

reason ______________________

IV. Spraying of Insecticide Plus Pesticide before Flowering

1. Are you aware about above practice	Yes-1 No-0

2. Have you sprayed before flowering	Yes-1 No-0
3. How many sprayings you have taken	Three-3 Two-2 One-1
3. For how much area	Whole-3 More than half-2 Less than half-1
4. When did you spray	After seeing symptoms Which may result in economic loss-3 Before any disease or pest attack-2 After full attack-1
5. Rate per acre	Recommended dose-3 Above recommendation-2 Below recommendation-1

6. If not adopted or not adopted as such what modified or alternative measures you have taken, specify reason.

Alternative/modification____________________
reason ____________________

V. Spraying of Insecticide Plus Pesticide after Flowering

1. Are you aware about above practice	Yes-1 No-0
2. Have you sprayed after flowering	Yes-1 No-0
How many sprayings you have taken	Three-3 Two-2 One- 1
3. For how much area	Whole-3 More than half-2 Less than half-1

4. When did you spray arter chilli harvesting	Within 1-2 days-3 Within 3-4 days-2 Within 5-6 days-1
5. Rate per acre	Recommended dose-3 Above recommendation-2 Below recommendation-1

6. If not adopted or adopted as such what modified or alternative measures you have taken, specify reason

Alternative/modification______________________
reason ______________________

VI. Application of Water Soluble Sulphur

1. Are you aware about the above practice	Yes-1, No-0
2. Did you spray with water soluble sulphur	Yes-1, No-0
3. For how much area	Whole -3 More than half-2 Less than half-1
4. When did you spray	After seeing attack which may result in economic loss-3 Before disease attack-2 After full attack-1
5. Rate per acre	Recommended dose-3 Above recommendation-2 Below recommendation-1

6. If not adopted or not adopted as such, what modified or alternative measure you have taken, specify reasons

Alternative/modification______________________
reason ______________________

Total adoption score

13. Plant Protection Status Index

The following area some of the statements regarding knowledge, skill, attitude and adoption of plant protection. You required to go through each statement and state your reaction about statements in terms of "strong agreement" “partial agreement”and “ disagreement”

(SA- Strongly agree, PA-partial agree, DA-Disagree)

Sr. No.	*Statements*	*Score SA PA DA*

(A) Knowledge

1. Farmer has knowledge about the availability of insecticide.
2. A farmer has knowledge about compatibility of various insecticides.
3. A farmer knows per hectare dose of spray as per crop growth.
4. A farmer has knowledge about the use of sprayers and dusters.
5. A farmer has knowledge about the nature of damage due to pest.
6. A farmer has knowledge about method of preparation of correct concentration of insecticides for spraying
7. A farmer has knowledge about selection of proper insecticides by identifying the pest.
8. A farmer has knowledge about the efficiency of insecticides against pest and effect of insecticides on parasites and predators.

Sr. No.	*Statements*	*Score SA PA DA*
9.	A farmer has knowledge about relationship of the date of sowing and infestation of pest.	
10.	A farmer has knowledge about classification of insecticides.	
11.	A farmer has knowledge about scientific method of preparing insecticidal spray for spraying the crop.	
12.	A farmer has knowledge about the centres for major repairs of sprayers and dusters.	
13.	A farmer has knowledge about the life cycle of insect-pests of crop.	
14.	A farmer has knowledge about non-utilization of salty water in preparation of insecticidal spray.	
15.	A farmer has knowledge that due to rotation of crops, infestation of insect is reduced.	
16.	A farmer has knowledge about the alternate host plants of the pest.	
17.	A farmer has knowledge about the burning of previous years' plant residues for checking the infestation of the pest in future.	
18.	A farmer has knowledge about the correct time of spraying or dusting of insecticide.	

Sr. No.	*Statements*	*Score SA PA DA*
19.	A farmer has knowledge about the appropriate insecticides and their dosages with their proper uses	
20.	A farmer has knowledge about the relation between the pest infestation and ecological conditions.	
21.	A farmer has knowledge about the identification of insects and diseases.	
22.	A farmer has knowledge about the difference between various insecticides of different companies and their relative costs.	
23.	A farmer has knowledge about the insecticides and fungicides.	
24.	A farmer has knowledge about the various information media about plant protection.	
25.	A farmer has knowledge to decide the correct dosages of insecticides.	
26.	A farmer has knowledge about the hazards of insecticide and its remedial first aid.	
27.	A farmer has knowledge about the respray of insecticide if rains are received within eight hours after spraying.	
28.	A farmer has knowledge about the life cycle of the pest and its stage of damaging the crops.	
29.	A farmer has knowledge about the thorough spraying and dusting of the crop.	

Sr. No.	*Statements*	*Score SA PA DA*
30.	A farmer has knowledge about the maintenance of sprayers and dusters after use.	
31.	A farmer has knowledge about the percentage of technical material in insecticides available in the market.	
32.	A farmer has knowledge about the characteristics of insecticides and fungicides.	
33.	A farmer has knowledge about the expiry date of insecticide to avoid its purchase.	
34.	A farmer has knowledge about the proper interval of spraying/ dusting.	
(B) Skill		
1.	A farmer has skill in deciding proper dose of insecticide.	
2.	A farmer has skill to decide as to when spraying or dusting of insecticide is to be started.	
3.	A farmer has skill in deciding the direction of spraying or dusting in relation to the direction of wind movement.	
4.	A farmer has skill of identifying the pest and applying proper insecticide	
5.	A farmer has skill in preparing proper insecticidal spray by knowing the technical material in the product.	

Sr. No.	*Statements*	*Score SA PA DA*
6.	A farmer has skill in identifying accurate insect pests and appropriate insecticide.	
7.	A farmer has skill in identifying important pest of cotton crop.	
8.	A farmer has skill in using appropriate insecticide in appropriate dosages.	
9.	A farmer has skill in preparing proper concentration of insecticides.	
10.	A farmer has skill in maintenance and repairs of sprayer/duster.	
(C)	**Attitude**	
1.	Various sprayers/dusters and their spare parts available in the market are satisfactory.	
2.	It is better to hire sprayer/duster than having our own.	
3.	Position of availability of insecticide is satisfactory in the market.	
4.	Plant protection has unique importance in increasing yield per hectare and improving the quality of produce.	
5.	The quality of insecticides available in the market is reliable.	
6.	Farmers are doubtful about the diagnosis of pest and recommendations made by the Extension workers from Agriculture Department, Zilla Parishad and other organization.	
7.	Rational use of insecticides is definitely beneficial.	

Sr. No.	*Statements*	*Score SA PA DA*
8.	A skillful crop rotation can avoid the damage from the pests to some extent to crop.	
9.	The insecticides available with Government organizations are of standard quality with reasonable price than the insecticides available in the market.	
10.	The spraying carried out by the same worker is more effective.	
(D)	**Adoption**	
1.	Spraying and dusting of the insecticides is done by considering the availability of important inputs.	
2.	The principle of crop rotation is always followed by farmer.	
3.	Insecticides should always be used by measuring them properly.	
4.	Spraying/dusting is carried out by considering the wind direction.	
5.	Proper dose of insecticide is used.	
6.	While handling the insecticides they should not be allowed to come in contact with wounds/mouth/nose.	
7.	Farmer is having his own sprayer/duster.	
8.	The plant protection recommendations given by Agriculture Department should be followed as such.	
9.	By considering the availability of insecticides in the market, farmer purchases the insecticides well in advance.	

Sr. No.	Statements	Score SA PA DA
10.	By considering the compatability of insecticides, tank-spray of insecticide is prepared.	
11.	After using sprayer/duster it is maintained properly.	
12.	If there are rains within eight hours from the spraying/dusting, respray/dusting is done.	
13.	The plant residues of the last year are destroyed before sowing another crop.	
14.	Spraying/dusting is done from all the sides of the plant.	
15.	Insecticidal spray is prepared by scientific method.	
16.	Farmer uses the skill in identifying the insects and diseases before undertaking plant protection measures.	
17.	After identifying the pest, proper insecticide is used in proper dosages.	
18.	While purchasing insecticide due consideration is given to efffectiveness and price of insecticides.	
19.	Uses appropriate insecticides by taking into account the climatic variations.	
20.	The sowing is done at appropriate time for the control of certain pest.	
21.	While spraying/dusting appropriate protective measures are taken and after operation the workers taken clear bath by using soap.	
	Total Score	

14. Problems

What are the various problems in the following areas you have been encountered/encountering in the adoption of various practices of plant protection measures.

I.	Situational	-1, 2, 3, 4, 5, 6, 7
II.	Economical	-1, 2, 3, 4, 5, 6, 7
III.	Technological	-1, 2, 3, 4, 5, 6, 7
IV.	Technical problems at personal level	-1, 2, 3, 4, 5, 6, 7
V.	Supply	-1, 2, 3, 4, 5, 6, 7
VI.	Technical guidance and services	-1, 2, 3, 4, 5, 6, 7

15. Suggestions

With your experience, what are your suggestions to bring about further improvement in the field so that more and more farmers can be brought into the adoption of plant protection technology in the dryland agriculture.

APPENDIX-II

PATH ANALYSIS

Path coefficients Showing direct and indirect effects of the independent variables on the PPS of DLFs

TABLE OF PATH COEFFICIENT DIRECT AND INDIRECT EFFECTS

Variable No.	*Name of the variables*	X_{14}	X_{13}	X_{12}	X_{11}	X_{10}	X_{09}	X_{08}	X_{07}	X_{06}	X_{05}	X_{04}	X_{03}	X_{02}	X_{01}
X_{14}	Adoption	-0.0214	0.6594	0.3038	-0.2255	0.0255	0.0155	-0.0380	0.0898	-0.3383	0.0466	0.1727	-0.0156	0.1780	0.0060
X_{13}	Utilization of source of information	-0.0175	0.8045	0.3057	-0.2456	0.0297	-0.0566	0.1165	-0.1215	-0.4012	0.0537	0.02005	0.0150	0.1839	0.0049
X_{12}	Risk preference	- 0.0140	0.5311	0.3075	0.0385	-0.542	0.0981	-0.0769	-0.2643	0.069	0.0469	0.1238	-0.0173	0.1653	0.0010
X_{11}	Scientific orientation	-0.0140	0.5744	0.4139	0.3440	0.0384	0.0534	0.1058	0.0900	-0.2617	0.0500	0.1293	-0.0100	0.1732	0.0046
X_{10}	Economic motivation	-0.121	0.5280	0.3942	-0.2922	0.0452	-0.0564	0.0952	-0.0935	-0.2996	0.0562	0.1355	-0.0098	0.1830	0.0035
X_9	Cropping intensity	-0.0068	0.3784	0.2085	-0.1526	0.0212	-0.1203	0.1130	-0.0640	-0.2653	0.0468	0.1553	-0.123	0.1530	0.0029
X_8	Agri. infra.	-0.0127	0.4655	0.2256	-0.1806	0.0214	-0.0675	0.2014	-0.0926	-0.3050	0.0450	0.1950	-0.0149	0.1243	0.0026
X_7	Avail. of PP equipment	-0.0099	0.5064	0.1845	-0.1604	0.0219	-0.0399	0.0966	-0.1930	-0.2851	0.0407	0.1508	-0.0106	0.1162	0.0068
X_6	Socio-economic status	-0.0146	0.6529	0.2476	-0.1821	0.0274	-0.0646	0.1243	-0.1113	-0.4944	0.0662	0.2537	-0.0195	0.2051	0.0086
X_5	Soc. parti.	-0.0111	0.4315	0.2418	-0.1919	0.0283	-0.0628	0.1009	-0.0875	-0.3649	0.0897	0.1567	-0.0110	0.1995	0.0039
X_4	Land holding	-0.0124	0.5422	0.1927	-0.1495	0.0206	-0.0628	0.1320	-0.0979	-0.4216	0.0473	0.2975	-0.0183	0.1525	0.0064
X_3	Family size	-0.0069	0.2483	0.1650	-0.0708	0.0091	-0.0304	0.0617	-0.0420	-0.1979	0.0203	0.1121	-0.0486	0.0274	-0.0063
X_2	Education	-0.0132	0.5153	0.2666	-0.2075	0.0288	-0.0641	0.0872	-0.0781	-0.3532	0.0624	0.1580	-0.0046	0.2871	0.0135
X_1	Age	0.044	-0.1337	-0.0151	0.0541	-0.0054	0.0120	-0.0179	0.0447	0.1444	-0.0120	-0.0645	-0.0103	-0.1319	-0.0294

INDIRECT EFFECTS

REFERENCES

1. Ambalagan, S. (1974) : *A study of factors influencing adoption of package of practices for high yielding varieties of paddy.* unpublished. M. Sc. Thesis, Agricultural College and Research Institute, Tamil Nadu, Coimbatore.
2. Anonymous (1986) : *Report of the Evaluation of Committee on Lab to Land programme,* Indian Council of Agricultural Research, New Delhi, p 78.
3. Appa Rao, G. *et al.* (1980) : Interpersonal relation indiffusion of an innovation in two Indian villages. *Indian Journal of Extension Education,* Vol. 16 No. 1 and 2, pp 19-24.
4. Balkishnan Singh, *et al.* (1985) : Some selected socio personal and economic characteristics of farmers of progressive and Non-progressive dairy village, *IJEE,* Vol. 21 No. 3 and 4, pp 101-103.
5. Bhaskaram, K. (1970) : Diffusion and adoption of agricultural innovations in Research and Extension Education for accelerating development process. (Ed.) Singh, K. N., *Indian Society of Extension Education,* New Delhi, pp 356-358.
6. Bhat, M. V. (1985) : Introduction to pesticides, *Pesticides Information,* Vol. 11 No. 3, pp 20-31.
7. Bhatnagar, O. P. (1986) : Experience in Technology transfer and Training at National Institute of Rural Development, Hyderabad *(Mimeo.).*
8. Bunting, A. M. (1981) : Preface, Management of Transfer of Farm Technology, NIRD, Hyderabad, pp V to VI *(Mimeo.).*
9. Byra Reddy, H. N. (1971) : *A study of differential characteristics of adopters and non-adopters of fertilizers to rainfed ragi* in Bangalore North Taluka, unpublished M. Sc. Thesis, Department of Agriculture Extension, University of Agricultural Sciences, Bangalore.

10. Chauhan, *et al.* (1973) : *Small Farmers' Problems and Possibilities of Development : A Study of Maharashtra State.* I. I. M., Ahmedabad.

11. Chinnamani, *et al.* (1970) : Modern dry farming methods, *Intensive Agriculture,* 8(6) pp 6-8.

12. Dak, T. M. and S. S. Purohit (1978) : A study of technological information gaps and Developing a Training strategy for DPAP *Areas of Haryana,* Extension Education Institute, Nilokheri (Mimeo.).

13. Desai, G. R. (1977) : Impact of block demonstration in participant and non-participant farmers of Dharwad District of Karnataka *State.* Unpublished M. Sc. Thesis, Division of Agricultural Extension, College of Agriculture, Dharwad.

14. Dhanakumar, V. G. (1981) : *An evaluation of the beneficiaries of Integrated Dryland Agricultural Development Project in respect of Dryland Technology,* Department of Agricultural Extension and Rural Sociology Centre for Agriculture and Rural Development Studies, Madurai.

15. Dhanorkar, T. S. (1970) : *Differential perception of farmers about the attributes of farm innovations.* M. Sc. (Thesis) Unpublished, College of Agriculture, Nagpur, pp 36-39.

16. Doijad, P. S. (1973) : *A study of village local leaders in relation to their adoption of agricultural practices and role performance,* Unpublished M. Sc. (Agri.) Thesis, College of Agriculture,Nagpur.

17. Doijad, P. S and D. Sen (1981) : Factors related to adoption behaviour of village local leaders, *Psychological Research Journal,* Vol. 5 No. 1 and 2. Psychological Research Academy suite, 9, 37, Syed Amir Ali Avenue, Calcutta, pp 39-46.

18. Dwarikanath, R. and Channe Gowda (1974) : Constraints to the transfer of new farm technology In *Transfer Technology Systems and Constraints,* Edited by D. Singh and J. S. Pandey, Nirbal-ke Balram Press, Kanpur, p 35.

19. FAO (1985) : Pesticides Kill one farmer every hour. *Newstime,* Daily, 6th September, 1985.

20. Ganasekaran, K. (1978) : *Information seeking and adoption behaviour of Neo-marginal Farmers,* Unpublished M. Sc. (Agri.) Thesis, Agriculture College and Research Institute, Tamil Nadu Agricultural University, Coimbatore.

21. Gartrell, J. W. and C. D. Gartrell (1979): Status, knowledge and Innovation. *Rural Sociology* 44: pp 73-91.

22. Gomez (1977) : *Multiple Cropping: An Approach to Rural Development :* College, Laguna, Philippines, University of Philippines at Losbanos O. P. D.

23. Gopalkrishna, G. (1978) : *A study on comparative effectiveness of different Farmers' Training programmes in A. P.* Unpublished M. Sc. Thesis, Andhra Pradesh Agricultural University, Hyderabad.

24. Government of Bihar (1985) : *Report on Monitoring M and E Unit-cum-Evaluation Survey (1984-85),* Directorate of Agriculture, Patna, Bihar.

25. Hans, *et al.* (1979) : *The nature and significant of risk in the SAT, socio-economic constraints to development of SAT Agriculture,* International Crop Research Institute for Semi Arid Tropics, Patancheru, Hyderabad p. 303.

26. Jaiswal, N. K. (1965) : *A study of factors associated with low level of adoption of improved agricultural practices.* Unpublished Ph.D. Thesis, Division of Agricultural Extension, Indian Agricultural Research Institute, New Delhi.

27. Jaiswal, N. K. and N. P. Singh (1968): Farmers, response to innovation in farming. A study in two East Bihar villages. *Journal of Agricultural Research,* Bhagalpur University, Vol. 1: pp. 34-36.

28. Jaiswal, N. K. (1970) : Effects of size of holding and irrigation facilities on adoption of HYV. *Indian Jouranl of Extension Education,* 6 : pp. 8-11.

29. Jaiswal, N. K. and H.P.S. Arya (1979) : *Management of Transfer of Farm Technology,* National Institute of Rural Development, Hyderabad.

30. Jaiswal, N. K. and A. P. Purandare (1982) : *Planning and Management of Watersheds in Solapur District,* National Institute of Rural Development, Hyderabad, p. 694.

31. Janakiramaraju, M. N. (1978) : *A study on the Fertilizer Adoption Behaviour of Farmers in Irrigated and Non-irrigated areas.* Unpublished M. Sc. (Ag.) Thesis, Agricultural College and Research Institute, Tamil Nadu Agricultural University, Coimbatore.

32. Jha, P. N. and S. N. Singh (1966) : Utilisation of sources of farm information as related to characteristics of farmers. *IJEE* 1 : pp. 294-362.

33. Jha, P. N. and G. S. Shaktwat (1972) : Adoption behaviour of farmers towards hybrid bajara cultivation. *Indian Journal of Extension Education.* 8 : pp 23-41.

34. Jock, R. Anderson (1979) : *The Nature and Significance of Risk Exploitation of New Technology Socio-economic Constraints to Development of Semi and Tropical Agriculture,* International Crops Research Institute for Semi Arid Tropics, Patencheru, pp. 297-302.

35. Jodha, N. S. (1979) : *Some Dimensions of Traditional Farming Systems in SAT*, India, International Crops Research Institute for Semi Arid Tropics, Patancheru, Hyderabad, pp. 11-24.

36. Kahlon, A. S. (1957) : *Farm Management Review Indian Council of Agricultural Research Technical Bulletin* (Agril.) No. 39, Indian Council of Agricultural Research, New Delhi.

37. Kalon, H. (1987) : The University of Byaffolo Studies 18: 23 cited by L. Gross (1975) *Social Theory: Inquiry and Paradigms*, Allied Publishers, Bombay, p.3.

38. Kulhari, V. S. (1980) : A study of reorganised Agricultural Extension. Training and Visit system and its impact in Chambal Command Area Development Project in Rajasthan. Unpublished Ph. D. Thesis, Indian Agricultural Research Institute, New Delhi.

39. Kunnal, *et al.* (1984): Adoption of new technologies in dryland Sorghum Crop production. *IJEE,* Vol. XX, No. 3 and 4, pp. 60-63.

40. Laxminarayana, K. (1982) : *Pest management in Sorghum Dryland Agricultural Research in India.* Thrust in Eighties, All India Co-ordinated Research Project for Dryland Agriculture, Hyderabad. pp.50-53.

41. Majumdar, A. H. and D. K. Majumdar (1967) : Adoption and some psychological characteristics of farmers. *Indian Journal of Extension Education.* 3. pp. 138-142.

42. Mann, B. P. S. (1985) : Welcome address at Annual General Meeting of PAI held at Delhi on 26th November, 1985. *Pesticides Information PAI,* New Delhi Vol. No. 3, p. 2.

43. Manivannan, N. (1980) : *A study on the Knowledge and Extent of Adoption of Sunflower Growers.* Unpublished M.Sc. (Agri.) Thesis, Agricultural College and Research Institute, Tamil Nadu Agricultural University, Coimbatore.

44. Mayani, V. V. and D. M. Sheth (1978) : Training of Small Farmers. *Kurukshetra,* 26 (12) pp. 21-22.

45. Mathizahagan, T. and R. P. Singh (1986): Training needs of banana growers. *IJEE,* Vol. 22 No. 1 and 2, pp 40-46.

46. Moulik, T. K (1965) : A study of the predictive values' of some factors of adoption of Nitrogenous fertilizers and the influence of sources of information on adoption behaviour. Unpublished Ph.D. Thesis, Division of Agricultural Extension, Indian Agricultural Research Institute, New Delhi.

47. Mruthyunjayam (1987) : Research on diffusion and adoption at Extension Education Institute, Rajendranagar, Hyderabad, (Abst.) *Souvenir* on 25 Years Silver Jubilee Celebrations,

Extension Education Institute, Andhra Pradesh Agricultural University, Hyderabad, p. 69-71.

48. Nagendranath, B. S. (1980) : Impact of Dryland Demonstrations conducted under Drought Prone Area Programme on the farmers of Chittoor, Andhra Pradesh Unpublished M. Sc. Thesis, Department of Extension Education, Sri Venkateswara Agriculture College, Andhra Pradesh Agricultural University, Tirupathi.

49. Narayana Rao, G. (1966) : A study on extent of knowledge and adoption of Hybrid Maize cultivation in Hazurabad Panchayat Samiti of Kareemnagar District. Unpublished Thesis, Extension Education, Institute, Rajendranagar, Hyderabad (A.P.)

50. Pandey, S. N. (1979) : *A study of communication patterns under the T and V system of Agricultural Extension in the Chambal Command Area Development Project of Rajasthan.* Unpublished Ph.D. Thesis, Indian Agricultural Research Institute, New Delhi.

51. Pandey, S.V. (1988) : Investigation on the analysis of pest control in cotton crop. Unpublished Ph. D. Thesis, Punjabrao Krishi Vidyapeeth, Akola.

52. Padheria, M. M. and I. C. Patel (1975) : Adoption pattern of recommended from practices in Gujarat. *IJEE.* Vol. II Nos.1 and 2, pp. 69-73.

53. Pathak, S. (1981) : A multiple regression analysis of factors related with attitude of farmers. *IJEE.* Vol. 17, Nos. 3 and 4 pp 73-80.

54. Pathak, S. and A. K. Majumdar (1978) : Multivariate Regression Analysis of adoption behaviour of farmers. *IJEE,* Vol. 14, Nos. 3 and 4, pp 45-50.

55. Pillai, Sadasivam (1978) : *A study on the Swine Farming and Pork Consumption,* Unpublished M. Sc. (Ag.) Thesis, Coimbatore Agricultural College and Research Institute, Tamil Nadu Agricultural University, Coimbatore.

56. Pradhan, S. (1983) : *Revolution in Pest Control, Agriculture Entomology Cheek and Pest Control,* Indian Council of Agricultural Research, New Delhi. p. 221.

57. Rai, H. N. (1967) : Diffusion of information and farmers' response relation to an improved new practice (Hybrid maize). *IJEE* .Vol. 1 : p. 240.

58. Raje, S. R. (1983) : *Operational Research: The Indo-UK Experience at Indore.* DLA Research in India, Thrust in the Eighties-All India Co-ordinated Research Project for Dryland Agriculture, Hyderabad pp 171-186.

59. Raj, Gopalan, C. and J. Singh (1971) : *Factors Hindering Adoption, Adoption of Agricultural Innovation*, National Publishing House, 23, Darya Ganj, Delhi-6, p. 99.

60. Raj Guru, O. S. and C. Satpathy (1971) : Effect of sources of information in the adoption process of HYV by the farmers of Puri District in Orissa, *IJEE*, Vol. 7 No. 1 and 2, pp. 71-75.

61. Ramesh, Babu and B. P. Sinha (1985) : Communication Behaviour of farmers with respect to modern rice technology, *IJEE* 3 (1-2) pp 51-58.

62. Randhawa, N. S. (1989) : Research has raised from productivity, *Hindustan Times*, Daily 19th January, 1989.

63. Rao, C. S. S. and T. K. Moulik (1966) : Influence of information sources, *IJEE* 2: pp. 43-48.

64. Rao, T. R. (1972) : A study on the traits associated with users and non-users of farm machinery and problems encountered in Farm Mechanisation. Unpublished M. Sc. (Ag.) Thesis, Agril. College and Research Institute, Tamil Nadu Agricultural University, Coimbatore.

65. Rao, Sarveswar (1985) : *Manufacturing and sale of improved Agricultural implements and their popularisation.* Indian Society of Agricultural Engineers, A. P. Chapter.

66. Rao, M. V. (1988) : Crop pests worry ICAR, *The Tribune,* (Daily) 18th June, 1988.

67. Rastogi, B. K. (1979) : *Cropping Patterns, Farming Practices and Economics of Major Crops in Selected Dryland Farming Regions of India.* International Crops Research Institute for Semi Arid Tropics, pp. 25 and 26.

68. Reed *et al.* (1980) : *Pest Management in Low-input Pigeonpea International Workshop on Pigeonpea pea.* Vol. 1, International Crops Research Institute for Semi Arid Tropics, pp. 99-105.

69. Reddy, M. S. K. (1962) : *A study of adoption of improved agricultural practices socio-economic factors and sources of informtion.* Unpublished M. Sc. thesis, Division of Agricultural Extension, Indian Agricultural Research Institute, New Delhi.

70. Reddy, Laxmi, B. and C. Venku Reddy (1983) : *Transfer of DAT and Behavioural Changes among the Farmers* (Abst.) Research Studies in Extension Education, Extension Education Institute, Andhra Pradesh Agricultural University, Rajendranagar, Hyderabad-500030 (A. P.), pp. 31-36.

71. Rogers, E. M. (1962) : *Diffusion of Innovation* (Third edition). The Free Press, Collier, Macmillan Publishers, London.

72. Sen, D. (1975) : *Symbolic Adopters and Agricultural Infrastructures : A Study on the Farmers of Some Selected Villages of S. Bengal.*Ph. D. Thesis, Department of Agricultural Extension, B. C., K. V. V. Kalyani, West Bengal.

73. Sen, D. and P. S. Doijad (1982) : Influence of utilisation of sources of information on the adoption behaviour and productivity of farmers of progressive and non-progressive villages; *Man and Life.* Institute of Social Research and Applied Anthropology, Vol. 8 No. 1 and 2 pp. 39-47.

74. Sen, D and P. S. Doijad (1983) : Factors influencing the Adoption tivity, *Pesticides,* January, 1983 pp. 3-7.

75. Sharma, S. K. and C. T. Nair (1974) : A multivariate study of high yielding varieties of paddy. *Indian Journal of Extension Education,* 10 (1 and 2): pp. 30-39.

76. Singh, B. N. and P. N. Jha (1965): Information in relation to adoption of improved Agriculture practices. *IJEE 1* : pp 34-42.

77. Singh, K. N. and C. Shankariah (1968) : Measuring the information source credibility with paired comparison technique. *Behavioural Sciences and C. D. 2* : pp. 38-44.

78. Singh, D. and M. R. Singh (1972): Farmers perception of different Credibility pattern of information sources. *IJEE,* 8 (3, 4) pp. 74-78.

79. Singh, A. D. (1974) : *Impact of Farmers' training on knowledge attitudes and adoption with respect to HYVS of wheat in Gwalior District of M. P.* Unpublished M.Sc. Thesis, Extension Education Institute, Andhra Pradesh Agricultural University, Hyderabad.

80. Singh, T. R. (1975) : *Analysis of Rationality in Decision-Making in relation to the Adoption of innovation of varying nature.* Unpublished Ph. D. Thesis, Division of Agricultural Extension, Indian Agricultural Research Institute, New Delhi.

81. Singh, Buta (1985) : Inaugural address by the Chief Guest S. Buta Singh, Union Minister of Agriculture. *Pesticides Information.* Vol. 11 No. 3 pp. 9-11.

82. Sinha, P. R. R. (1966) : A study of farmers decision-making process with respect to use of artificial fertilizers and vegetable cultivation in Krishwala Block, Delhi. Unpublished Ph.D. Thesis, Division of Agricultural Extension, Indian Agricultural Research Institute, New Delhi.

83. Sinha, B. P. (1974) : *Television in Diffusion of Farm Information : A Motivational Study).* Indian Printers, Delhi-6.

84. Sinha, *et al.* (1976) : Training needs of small and marginal farmers. Paper presented at National Seminar on New Agricultural Technology and Extension Strategy for small and marginal Farmers held at Punjab Agricultural University, Ludhiana.

85. Sinha, *et al.* (1984) : A study of attitude of farmers of Chhota Nagpur towards soil conservation programme. *IJEE.* Vol. 20 (3 and 4), pp. 9-18.

86. Sinha, *et al.* (1988): Communication behaviour of opinion leaders of Nepal. *Indian Journal of Rural Development,* Vol. 7 (3), National Institute of Rural Development, Hyderabad. p. 336.

87. Solane (1978) : Factors affecting inputs used and farm productity on different farm categories in Punjab. *Pakistan Development Review,* 17 (3) : pp. 316-332.

88. Somasundram, D. (1976) : A diagnostic study of small farmers with respect to New Agricultural Technology and its effective communication for adoption. Unpublished Ph. D. Thesis, Division of Agricultural Extension, Indian Agricultural Research Institute, New Delhi.

89. Srinivasa Murthy, J. (1965): What visual aids were employed to teach farmers in the use of fertilizers in *Dryland Ragi Souvenir,* Alumni Association, College of Agriculture, Bangalore.

90. Subramanyan, V. S. and K. B. Sripaul (1976): A study of factors influencing farmers in the adoption of recommended farm practices for Millets (Cholam and Cumbu) in *Irrigated and Dryland Research Project No. R. 1 Ag Extn. 203,* Dept. of Agrl. Extension, Agrl. College and Res. Instt. TNAU, Coimbatore.

91. Sudarshnan Rao, *et al.* (1983): *Effectiveness of IAEP (T and V) in A. P. (Abst.)* Research Studies in Extension Education, Ph.D. Thesis 1980-82. Extension Education Institute, Andhra Pradesh Agricultural University, Hyderabad, pp. 52-61.

92. Sukumaran, N. (1972) : *Impact of Farmers' Training on knowledge and adoption of high yielding varieties of paddy in Trichur District of Kerala State,* Unpublished M.Sc. (Agri.) Thesis, Extension Education Institute, Andhra Pradesh Agricultural University, Hyderabad.

93. Sundaraswamy, B. (1971) : Extent of Adoption of recommended Practices and information sources consulted by the farmers in respect of Hybrid-jowar cultivation in the selected taluk of Mysore District, Unpublished M.Sc. (Ag.)

Thesis, Agricultural College and Research Institute, Tamil Nadu Agricultural University, Coimbatore.

94. Supe, S.V. (1969) : Factors related to different degrees of rationality in decision-making among farmers. Unpublished Ph.D. Thesis, Division of Agricultural Extension, Indian Agricultural Research Institute, New Delhi.

95. Supe, S.V. and M. S. Salode (1975) : Impact of national demonstration on knowledge and adoption level of farmers' participants. *IJEE*. Vol. 11, *PP*. 36-39.

96. Trigunayat, K. P. (1971) : Impact of radio and farmers, training and education programme in U.P., Unpublished M. Sc. Thesis, Extension Education Institute, Andhra Pradesh Agricultural University, Hyderabad.

97. Veerabhadraiah, V. and B. Dwarakinath (1970) : A study of adoption of Hybrid Maize in Bangalore District, Mysore Journal of Agricultural Sciences, 4 (3): pp. 326-336.

98. Venkateswarlu, J. (1987) : An era of self-sufficiency, Indian Council of Agricultural Research, New Delhi, p. 174.

99. Venkatramaiaha, P. (1983) : Development of socio-economic status scale for farm families of North Karnataka. Unpublished Ph.D. Thesis, Department of Extension, University of Agricultural Sciences, Bangalore.

100. Venkiduswamy, G. (1977) : Study of the Motives of small farmers in Project (SFDA) and Non-project Blocks for the adoption of High Yielding varieties of Cotton, Unpublished M. Sc. (Ag.) Thesis, Agricultural College and Research Institute, Tamil Nadu Agricultural University, Coimbatore.

101. Vijayaraghavan, R. (1979) : Critical analysis of functioning of Integrated Dryland Agriculture Development Project. Unpublished M. Sc (Ag.) Thesis, Agricultural College and Research Institute, Tamil Nadu Agricultural University, Coimbatore.

102. Vittal, M. (1971) : A study of production-cum-Demonstration training camps organised under Farmers' Training programme in North Arcot District of Tamil Nadu. Unpublished M. Sc. Thesis, Extension Education Institute, Andhra Pradesh Agricultural University, Hyderabad.

INDEX